Mastering Blockchain Programming with Solidity

Write production-ready smart contracts for Ethereum blockchain with Solidity

Jitendra Chittoda

BIRMINGHAM - MUMBAI

Mastering Blockchain Programming with Solidity

Commissioning Editor: Richa Tripathi
Acquisition Editor: Shriram Shekhar
Content Development Editor: Manjusha Mantri
Senior Editor: Afshaan Khan
Technical Editor: Pradeep Sahu
Copy Editor: Safis Editing
Project Coordinator: Prajakta Naik
Proofreader: Safis Editing
Indexer: Rekha Nair
Production Designer: Shraddha Falebhai

First published: August 2019

Production reference: 1010819

Published by Packt Publishing Ltd.
Livery Place
35 Livery Street
Birmingham
B3 2PB, UK.

ISBN 978-1-83921-826-2

www.packtpub.com

I dedicate this book to my wife Neha, and my two daughters Arshiya and Advika with much love.

Packt.com

Subscribe to our online digital library for full access to over 7,000 books and videos, as well as industry leading tools to help you plan your personal development and advance your career. For more information, please visit our website.

Why subscribe?

- Spend less time learning and more time coding with practical eBooks and Videos from over 4,000 industry professionals

- Improve your learning with Skill Plans built especially for you

- Get a free eBook or video every month

- Fully searchable for easy access to vital information

- Copy and paste, print, and bookmark content

Did you know that Packt offers eBook versions of every book published, with PDF and ePub files available? You can upgrade to the eBook version at www.packt.com and as a print book customer, you are entitled to a discount on the eBook copy. Get in touch with us at customercare@packtpub.com for more details.

At www.packt.com, you can also read a collection of free technical articles, sign up for a range of free newsletters, and receive exclusive discounts and offers on Packt books and eBooks.

Foreword

I have known Jitendra Chittoda for a good while. Together, we have audited a significant number of smart contracts for different blockchain platforms. Some of these contracts have been widely used, while others handle large amounts of funds.

Prior to working as a smart contract auditor, Jitendra was working as a developer, which gave him a broader perspective of the field, as he has observed it from a variety of standpoints. These multiple viewpoints of blockchain, and Ethereum in particular, have allowed him more general insights.

While many have heard about the power and potential of blockchain and smart contracts, I believe the blockchain journey is still in its infancy. As with many new technologies, a number of mistakes have been made. Some mistakes led to exploits, and some exploits were so large that they made news headlines. One of those attacks was the DAO reentrancy attack.

Given the novelty of the technology, people require guidance. Such guidance is required on different levels. Firstly, many people still do not sufficiently understand what blockchain is really about, what makes it special, and what differentiates it from regular databases. They don't understand the power of smart contracts—which are sometimes called programmable money—and their potential impact on future businesses.

Secondly, while other people understand the power of blockchain, they do not see a clear path to building on top of it. They are not aware of the many small mistakes that can be made along the way, which range from overly expensive smart contracts to a complete loss of control. Hence, these people require detailed and explicit technical guidance.

Given Jitendra's background, he can provide a good overview and many technical recommendations that will not only speed up learning, but can also avoid costly and potentially project-threatening errors. This book starts off with introductions to blockchain before diving into Solidity, its details, its common applications, its most useful helpers, and its security practices.

Hubert Ritzdorf

Chief Technical Officer, ChainSecurity AG

Contributors

About the author

Jitendra Chittoda is a blockchain security engineer at ChainSecurity. His day job is to perform security audit on smart contracts and expose security vulnerabilities in Solidity and Scilla contracts. He has also developed a non-custodial, decentralized, P2P lending contracts for ETHLend. The Solidity contracts that he has developed or audited handle over $100 million worth of cryptoassets. He also served as a tech and security advisor in various ICO projects.

Before finding his passion for blockchain, he coded in Java for over 11 years. He is the founder and leader of Delhi-NCR-JUG, a non-profit meetup group for Java. He holds a master's degree in computer applications and is regularly invited as a speaker at various conferences and meetups.

I would like to thank my wife, Neha, for encouraging and supporting me every single day with writing this book. A big shout-out to Manjusha Mantri, Shriram Shekhar, Pradeep Sahu, all of the technical reviewers, and the Packt team for their peer review, suggestions, and full support. I would also like to thank all my mentors and professors who always helped me in becoming a programmer. Thank you, Mom and Dad. You have given me the greatest gift: an education.

About the reviewers

Natalie Chin is a blockchain developer, college professor, and technical blogger. She is currently teaching at George Brown College in Toronto, helping to lead the first blockchain college program in the world. At STK, she built a level-two scaling solution on Ethereum, allowing instant cryptocurrency purchases. Natalie won three prizes at ETHSanFrancisco, 2018, where her team, Lending Party, built a decentralized application allowing a CDP (cryptocurrency loan) to liquidate into a bank account instantly.

She is an avid hackathon organizer, associated with DeltaHacks and Stackathon, where she inspired new developers to join the blockchain space. Natalie is a Women in Tech ambassador and is always passionate about spreading knowledge about the blockchain field.

A sincere thank you to Dad, for introducing me to technology at a young age, supporting me, and perpetually pushing me outside of my comfort zone to grow and be better.

To Zichen Jiang, for always finding the silver lining of situations, and always inspiring me to embrace everything with wit, strength, and humor.

Finally, I am eternally grateful to my mother-in-law, for embodying what it means to live a life of curiosity and positivity.

Ben Weinberg is a self-taught computer programmer who started off in iOS development with Objective-C and Swift, and, eventually, got caught up in the world of blockchain while working in Jerusalem. He eventually started learning blockchain and web development simultaneously with the goal of entering the blockchain space as a developer. Ben worked at the Toronto-based Bitcoin Bay, writing Solidity smart contracts and leading workshops on Ethereum development and Bitcoin opcodes. He also joined George Brown College, where he assisted aspiring blockchain developers as a lab monitor for George Brown's new blockchain development course. At George Brown, he helped teach full-stack web development, smart contract development, and decentralized application development.

I would like to thank all the mentors who have helped me along the way in my endeavors to become a programmer. I would like to acknowledge the Toronto and Jerusalem blockchain communities for being so accessible and welcoming to me as I entered this world without knowing anyone and with little knowledge of how blockchain works. I would also like to send my most supreme thanks to my family for all they have done.

Marc Lijour helps organizations strengthen their competitive advantage with technology such as blockchain and practices such as DevOps. Marc brings consulting experience from the public sector, Cisco, Savoir-faire Linux, and ConsenSys to executives driving innovation. He started the Metamesh Group with 30 consultants spanning across 5 continents. He helped George Brown launch the first blockchain development college program in Canada.

Marc holds degrees in mathematics and computer science, as well as an MBA in technology and innovation. He serves on the board of several not-for-profit organizations, including the **Information and Communications Technology Council (ICTC)**, ColliderX, TechConnex, and the Toronto French Business Network.

Packt is searching for authors like you

If you're interested in becoming an author for Packt, please visit `authors.packtpub.com` and apply today. We have worked with thousands of developers and tech professionals, just like you, to help them share their insight with the global tech community. You can make a general application, apply for a specific hot topic that we are recruiting an author for, or submit your own idea.

Table of Contents

Section 4: Design Patterns and Best Practices

Preface

Blockchain technology is at its nascent stage. However, technology is slowly moving forward and new developments using blockchain are emerging. This technology has the power to replace trusted third parties with trusted blockchain networks. Bitcoin was the birth of blockchain technology and has shown the world a new peer-to-peer payment system without needing intermediaries. You could say that Bitcoin was the first generation of blockchain technology. However, Ethereum took this innovative technology to the next level—you could call it blockchain generation two—where you can write decentralized applications using smart contracts. Solidity is the most widely used and popular language for writing smart contracts for decentralized applications.

The book starts with explaining blockchain, Ethereum, and Solidity. It mostly focuses on writing production-ready smart contracts in the Solidity language for Ethereum blockchain. It covers basic Solidity language syntax and control structures, and moves on to writing your own contract. It also deep dives into different libraries that you can use while writing contracts. Later on, it covers tools and techniques to write secure, production-ready smart contracts.

Who this book is for

This book is for professional software developers who have started learning blockchain, Ethereum, and the Solidity language and who want to make a career in writing production-ready smart contracts. This book is also aimed at developers who are interested in building decentralized applications over Ethereum blockchain. This book will help you learn the Solidity language for building smart contracts from scratch and make you proficient in building production-ready smart contracts for decentralized applications.

What this book covers

Chapter 1, *Introduction to Blockchain*, starts with how blockchain technology came into existence through the innovation of Bitcoin. This chapter discusses the different properties of a blockchain. It also introduces Ethereum and how it is different from the Bitcoin blockchain. Later, this chapter introduces smart contracts.

Chapter 2, *Getting Started with Solidity*, starts with the basic Solidity language syntaxes and the structure of a contract. You will learn about the different data types available in Solidity. This chapter also discusses globally available variables present in the Solidity language that you can use while writing your smart contract.

Chapter 3, *Control Structures and Contracts*, deep dives into the control structures of Solidity contracts, the different types of functions supported, contract inheritance, and event logging.

Chapter 4, *Learning MetaMask and Remix*, discusses setting up your MetaMask plugin and creating wallets using it. This chapter also discusses using the online Remix IDE to create, compile, deploy, and interact with your Solidity contracts.

Chapter 5, *Using Ganache and the Truffle Framework*, discusses installing and setting up your local blockchain instance using Ganache. This chapter also discusses installing and setting up the Truffle framework, learning how to use its commands, setting up your new Truffle project, writing migration scripts for the Truffle framework, and adding test cases to a project.

Chapter 6, *Taking Advantage of Code Quality Tools*, discusses improving the quality of your contracts by using open source tools such as `surya`, which helps in generating different kinds of reports. This chapter also discusses using Solidity linters to lint your code and fix common issues present in your contracts, as well as running Solidity coverage tools to generate code coverage reports.

Chapter 7, *ERC20 Token Standard*, covers the introduction of the ERC20 token standard. This chapter deep dives into its full implementation details, provides an in-depth study of each function of the standard, and covers different events, optional functions, and a number of advanced functions.

Chapter 8, *ERC721 Non-Fungible Token Standard*, starts with an introduction to the ERC721 standard and the difference between the ERC20 and ERC721 standards. This chapter deep dives into each of the state variables, functions, and events associated with ERC721 implementation. This chapter also discusses some other optional contracts that can be used with the ERC721 standard.

Chapter 9, *Deep Dive into the OpenZeppelin Library*, starts with an introduction to OpenZeppelin library contracts. This chapter will help you learn how to install and use OpenZeppelin library contracts in your Truffle project. It also provides in-depth studies of library contracts such as `Ownable`, `Claimable`, `Roles`, `PauserRole`, `Pausable`, `ERC20`, `SafeERC20`, `DetailedERC20`, `ERC20Mintable`, `ERC20Burnable`, `ERC20Pausable`, `Math`, `SafeMath`, `Crowdsale`, `Address`, and `ReentrancyGuard`.

Chapter 10, *Using Multisig Wallets*, The multisig wallets are special contracts that require multiple signatures to execute a transaction. This chapter provides an introduction to multisig contracts and their usage. This chapter also covers installing, creating, and setting up your own multisig wallet, as well as how to use and control it.

Chapter 11, *Upgradable Contracts Using ZeppelinOS*, provides an introduction to the ZeppelinOS development platform. Topics covered include creating a new project using zos commands, creating an upgradable contract, deploying, re-deploying to upgrade the contract, and some precautions to take when using ZeppelinOS for writing upgradable contracts.

Chapter 12, *Building Your Own Token*, helps you learn how to create your own ERC20 token contract from scratch. You will learn how to draft the specification of a contract, set up a Truffle project, create a contract, choose which OpenZeppelin libraries to use, write migration scripts, write test cases, execute migration scripts and test cases in Ganache, and finally, test your contracts on a testnet such as Rinkeby.

Chapter 13, *Solidity Design Patterns*, introduces different Solidity design patterns. These are divided into five categories: *Security patterns*: Withdrawal, Access restriction, and Emergency stop; *Creational patterns*: Factory pattern; *Behavioral patterns*: State machine, Iterable map, Indexed map, Address list, and Subscription; *Gas economic patters*: String comparison and Tight variable packing; and *Life cycle patterns*: Mortal and Auto deprecate pattern.

Chapter 14, *Tips, Tricks, and Security Best Practices*, helps you to learn different Solidity smart contract best practices, such as avoiding floating pragma, the 15-second blocktime rule, rounding errors, and gas consumption. It also helps you to learn about different security attacks, such as front-running, reentrancy, signature replay, integer overflow and underflow, and how to prevent these attack patterns.

To get the most out of this book

You should have knowledge of any of the existing programming languages. Solidity language syntaxes are very similar to Java, JavaScript, and Python syntaxes. A developer who has created projects using Java, JavaScript, or Python can pick up and learn Solidity very easily. You should also have some basic knowledge of OOPS concepts, blockchain, wallets, private key, public key, consensus algorithms, and exchanges.

One very important thing to note here is that, in order to write smart contracts for decentralized applications, you need to have a completely new mindset. Up until now, you have probably been building applications using Java, JS, or another language in which you can fix bugs later on, even in the production environment when something goes wrong. However, smart contracts are immutable. If your contract is deployed in production, you are done. If there is a bug left in the contract, you will be unable to fix it. Hence, you need to ensure that all your smart contracts are well-tested and have no bugs present in them.

Download the example code files

You can download the example code files for this book from your account at www.packt.com. If you purchased this book elsewhere, you can visit www.packt.com/support and register to have the files emailed directly to you.

You can download the code files by following these steps:

1. Log in or register at www.packt.com.
2. Select the **SUPPORT** tab.
3. Click on **Code Downloads & Errata**.
4. Enter the name of the book in the **Search** box and follow the onscreen instructions.

Once the file is downloaded, please make sure that you unzip or extract the folder using the latest versions of the following:

- WinRAR/7-Zip for Windows
- Zipeg/iZip/UnRarX for Mac
- 7-Zip/PeaZip for Linux

The code bundle for the book is also hosted on GitHub at https://github.com/PacktPublishing/Mastering-Blockchain-Programming-with-Solidity. If there's an update to the code, it will be updated on the existing GitHub repository.

We also have other code bundles from our rich catalog of books and videos available at https://github.com/PacktPublishing/. Check them out!

Download the color images

We provide a PDF file that has color images of the screenshots/diagrams used in this book. You can download it at https://static.packt-cdn.com/downloads/9781839218262_ColorImages.pdf.

Code in action

To see the code being executed please visit the following link: http://bit.ly/2Yv6kpm.

Conventions used

There are a number of text conventions used throughout this book.

CodeInText: Indicates code words in text, database table names, folder names, filenames, file extensions, pathnames, dummy URLs, user input, and Twitter handles. Here is an example: "Solidity supports different data types, including uint, int, address, and many more."

A block of code is set as follows:

```
contract VariableStorage {
    uint storeUint; //uint256 storage variable
    //...
}
```

When we wish to draw your attention to a particular part of a code block, the relevant lines or items are set in bold:

```
contract VariableStorage {
    uint storeUint; //uint256 storage variable
    //...
}
```

Any command-line input or output is written as follows:

```
$ npm install -g ganache-cli
```

Bold: Indicates a new term, an important word, or words that you see on screen. For example, words in menus or dialog boxes appear in the text like this. Here is an example: "As mentioned earlier, you can start a local Ganache blockchain instance by just clicking on the **QUICKSTART** button."

 Warnings or important notes appear like this.

Tips and tricks appear like this.

Get in touch

Feedback from our readers is always welcome.

General feedback: If you have questions about any aspect of this book, mention the book title in the subject of your message and email us at customercare@packtpub.com.

Errata: Although we have taken every care to ensure the accuracy of our content, mistakes do happen. If you have found a mistake in this book, we would be grateful if you would report this to us. Please visit www.packt.com/submit-errata, select your book, click on the Errata Submission Form link, and enter the details.

Piracy: If you come across any illegal copies of our works in any form on the internet, we would be grateful if you would provide us with the location address or website name. Please contact us at copyright@packt.com with a link to the material.

If you are interested in becoming an author: If there is a topic that you have expertise in, and you are interested in either writing or contributing to a book, please visit authors.packtpub.com.

Reviews

Please leave a review. Once you have read and used this book, why not leave a review on the site that you purchased it from? Potential readers can then see and use your unbiased opinion to make purchase decisions, we at Packt can understand what you think about our products, and our authors can see your feedback on their book. Thank you!

For more information about Packt, please visit packt.com.

Section 1: Getting Started with Blockchain, Ethereum, and Solidity

In this section, the reader will learn about blockchain's what, why, and when. It will also include an overview of Ethereum blockchain, accounts, and transactions.

The following chapters will be covered in this section:

Introduction to Blockchain

1

Blockchain technology was born with the invention of Bitcoin—a new form of **peer-to-peer (P2P)** electronic cash—back in 2008. The Bitcoin white paper was released on October 31, 2008 and the first release of Bitcoin came on January 3, 2009. The blockchain space is still a toddler in terms of adoption and the tools available. It has some unique properties that did not exist before in any of the previous systems or software applications. The most important property of blockchain is establishing trust between two or multiple parties without needing any intermediaries, which opens a new era in programming.

Ethereum is one of the implementations of blockchain. Ethereum is an open source, public, and distributed computing platform. On Ethereum, developers can deploy smart contracts written in the Solidity language and build a decentralized application—also called dApp, Đapp, Dapp, or DApp.

Smart contracts are small programs where developers can define the rules of the trust that they intended to code. One of the mind-boggling properties of smart contracts is *immutability*—once they are deployed on the blockchain, their code cannot be changed. This immutable property makes it very hard to program smart contracts and predict errors/bugs beforehand.

In this chapter, we will learn more about blockchain technology, and when and where blockchains should be used. We will also introduce Ethereum and smart contracts.

The following topics will be covered in this chapter:

- Introduction to blockchain
- Properties of a blockchain
- What is Ethereum?
- Introduction to smart contracts
- Why smart contracts are different from traditional software programs

Understanding blockchain

A blockchain is a timestamped series of immutable transactions that is managed by a cluster of computers using special computer algorithms. These immutable records are not owned by any single entity. A blockchain is a decentralized P2P network of nodes. Each node in a blockchain shares the same copy of data, also called the **digital ledger**. Each node present in the network uses the same algorithm to reach a **consensus**.

A blockchain, by design, is resistant to the modification of data. The ledger can record transactions between two parties in a verifiable and permanent way. Whenever there is a change in the ledger using transactions, changes are distributed to all the nodes, to verify and update their own copy of the ledger. Once a transaction is stored and verified by all the nodes in the network, then it is not feasible to change the transaction without altering all the subsequent and previous blocks. That's why blockchain transactions are irreversible, as blockchain transactions and their data are append only.

Each computer that participates in this P2P network is called a **node**. Each node maintains the records of transactions in multiple consecutive blocks. The P2P network is also used in torrents such as BitTorrent; however, torrent networks are not like blockchains, as they are designed to shares files only.

Blockchain technology is also called **Decentralized Ledger Technology (DLT)**, as each node in the network keeps the same copy of the ledger. Please have a look at the following diagram:

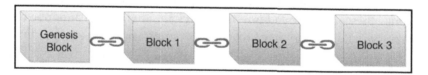

Chain of connected blocks

In the preceding diagram, each **block** is connected with a link (also known as a **chain**). The chain is usually recognized as the chain of all the blocks. The link between two blocks is implemented by having a record of the cryptographic hash of the previous block in each block, so that you can visit the chain in reverse chronological order.

One question that may arise in your mind is, what is the difference between a traditional software application that processes the transaction and a blockchain that also processes the transaction? Then, why you would need blockchain technology? Well, blockchain technology solved one of the hard problems of computer science—the **double-spending** problem. Let's understand what is the double-spending problem..

It is extremely hard to reverse a blockchain transaction. However, there have been cases when a 51% attack allowed an attacker to double-spend coins. One such example is when **Ethereum Classic** (symbol: **ETC**) was attacked with a 51% attack, in which approximately $1.1 million worth of ETC was lost.

Blockchain solves the double-spending problem

As we know, we can share documents and pictures over the internet with someone. When we share those documents or pictures with another person, we actually share a copy of the file. Once that copy is sent to another person, you and the other person have the same copy of that document.

However, things such as money, bonds, and shares must not be shared as copies, like we do for documents/pictures when we need to transfer them over the internet. If you try to send money P2P, without using any intermediaries, then both parties end up having the same copies of the money. This is called the double-spending problem.

For example, if Alice gives a $10 bill to Bob, then Alice must not hold a copy of the $10 bill, as it's given to Bob. These are called **value transfers**. To transfer those items that have values over the internet, we need to depend upon **Trusted Third Parties (TTPs)**. Banks and stock exchanges are the TTP on which we rely for transferring values (money and shares) from person A to B. For value transfers between two parties over the internet, they both have to depend upon a trusted middle party (centralized) to process the transactions and ensure the safety of transactions.

Blockchain solved this double-spending problem. Now, for value transfers between two parties, neither of them have to depend upon a middleman (trusted party).
They both can do safe transactions directly. Blockchain's decentralized network and consensus algorithm ensures the safety of transactions and prevents the double-spending problem.

Properties of blockchain

Blockchain has properties of both decentralized and distributed networks. Using those types of networks along with cryptography adds more properties. We are covering properties related to the Ethereum blockchain rather than other blockchain implementation properties. The other blockchain implementations might have different properties. Let's discuss those properties.

Distributed ledger

Multiple nodes make up a distributed blockchain network. All nodes share a common ledger, where records of transactions are kept.

Fault tolerance

A blockchain network is distributed and each node maintains the same record of the ledger. Even if some nodes in the network are corrupted or go down, it can continue operating safely up to a certain limit, as well as processing transactions with running nodes.

Attack resistance

A blockchain network does not have centralized control. The network's resistance to attacks is maintained by the miners who are putting their processing power (using nodes) into use to guard against malicious attacks. These miners earn some incentives to keep the network safe by behaving honestly. This is done by using the distributed network and cryptographic techniques.

Remove intermediaries

Blockchain technology removes the dependence on TTP/middle parties/intermediaries. Using blockchain technology, a transaction can be done directly between two entities/systems. In place of intermediaries, we can place blockchain systems.

Consensus protocol

This is a protocol that ensures that all nodes participating in the network ensure the safety of the network. All nodes use a consensus protocol (a specific algorithm) to reach a consensus and discard the blocks generated by the attacker/bad node, to avoid catastrophic system failure. For example, Bitcoin and the Ethereum blockchain, at the time of writing, work on the **Proof-of-Work (PoW)** consensus protocol. Ethereum uses the Ethash PoW consensus algorithm, which is **Application Specific Integrated Circuit (ASIC)** resistance. In the future, Ethereum will switch its consensus algorithm from PoW to **Proof-of-Stake (PoS)**.

The PoW protocol is a consensus algorithm in which nodes compete with other nodes to solve a cryptographic puzzle. The node that solves the puzzle first adds a new block to the blockchain (called mining). By doing this, they also earn some block reward in the blockchain's native coin.

The PoS protocol is a consensus algorithm in which a person has to put the native blockchain coin up for stake (lock). The protocol selects a miner randomly or according to coinage. The miner is rewarded when the block added is valid, otherwise they may lose their stake.

Faster settlement

Traditional banking systems can be slow in some cases, as they need additional time for processing a transaction; for example, cross-border payments.. However, with blockchain technology, there are no intermediaries; therefore, the transaction happens directly between entities and the settlement is much faster, compared to traditional systems.

Lower transaction fees

Using the traditional banking system, doing a cross-border payments is costly. The intermediaries take their commission to process the transaction between two parties. But by using blockchain technology, the cost of doing transactions is significantly lower because we can remove the intermediaries and perform the transactions directly.

Transparency

Some blockchain systems maintain transparency. Ethereum is a public blockchain network. All the transactions of the Ethereum blockchain are public and transparent. Anyone can see the balance and transaction history of any **wallet** at any time, just by accessing the Ethereum public blockchain via a block explorer. However, some work toward having privacy on the Ethereum blockchain is being done; for example, using the **AZTEC** protocol.

Immutability

Every transaction that happens on the Ethereum blockchain is immutable. Smart contracts on the Ethereum blockchain are also immutable. Once the smart contract code is deployed, it cannot be changed. Smart contract code will remain on the blockchain forever. Anyone can see any deployed smart contract any time in the future as well, just by putting its contract address on the block explorers. Data that is being stored in smart contract variables is also immutable, unless there are data removal techniques that have not been exposed by the contract code.

Irreversible transactions

Once a transaction is executed on a blockchain and it receives sufficient confirmation, the transaction becomes irreversible. The irreversible transaction ensures the safety of the value transfer. The higher the number of confirmations received for a transaction, the harder it becomes to reverse the transaction from the attacker.

Trust in the network

As of today, we are using the existing TTP systems of bank servers or stock exchange trade servers. These are centralized and are continuously being attacked. If an attacker takes control of the database server of a trade server, it can make changes in the database and put in fake entries.

However, the immutability and irreversibility of transactions makes blockchain technology more trusted. It becomes very hard for an attacker to change any entries in the blockchain transactions. This makes blockchain more trusted in terms of maintaining pure immutability and irreversible transactions. You can trust the Ethereum public mainnet; however, you cannot trust testnet completely (this may change, but at the moment, testnet is only used for testing).

I am going to define two types of trust here—**artificial trust** and **trustless systems**.

Artificial trust

When replacing trusted middle parties (intermediaries) with blockchain technology, I call it artificial trust. Blockchain technology will be trusted while doing P2P transactions. A person only has to trust the blockchain technology that is going to execute their transaction.

Trustless systems

Two people, without knowing each other, can do transactions using blockchain technology. They do not have to trust each other, they only have to trust blockchain technology. For both persons, it is a trustless system.

Availability

Centralized systems sometimes have to go through regular maintenance and downtime, resulting in inconvenience to the users. The blockchain is decentralized, so even if some nodes go down/are corrupted, the network will continue to function and will be available to process transactions 24/7.

Empower individuals

Each individual entity/person has their own wallet's public and private keys on the blockchain network. Using those wallets, they are in full control of their assets and the privileges available for those wallets. Blockchain assures that the ownership of a person's data is in their hands.

However, an individual can only maintain full control over their cryptocurrency or data when they own the private keys of their wallet. When the private keys of their wallet is maintained by another entity, such as centralized exchanges, they may not have full control.

 Wallet: A wallet is software that keeps one or more cryptographic private and public keys. Using these keys, you can interact with different blockchains and are allowed to send and receive digital currencies. You can also interact with smart contracts using any of the accounts present in your wallet.

Chronological order of transactions

Blockchains keep their transactions in blocks. Multiple transactions are coupled together and stored in a new block. Next, that new block is appended to the chain of previous blocks. This keeps all the blockchain transactions maintained in chronological order.

Timestamped

Every transaction on the blockchain is stamped with the current time. This enables the blockchain to maintain all the transaction histories of an account. Also, this can be used to prove whether or not a transaction happened on a certain date and at a certain time.

Sealed with cryptography

Asymmetric cryptography is used for wallet generation and transaction signing using the **Elliptic Curve Digital Signature Algorithm (ECDSA)**. All transactions are packed together using the Merkle tree and SHA-256 cryptographic algorithms. This makes blockchains secure and reliable to use.

When to use blockchain

As we have looked into the double-spending problem, blockchain technology solves this problem and opens new opportunities to create programs/systems where we can remove the trusted parties and make value transfers between two people directly.

The first implementation of blockchain is the Bitcoin blockchain. The Bitcoin blockchain solves the problem of making value transfers. A person can transfer bitcoins (these have some value) to another person without needing a middle party to process the transaction. Because *trust* is now transferred from middle parties (TTP) to the blockchain decentralized network, which uses a consensus algorithm called PoW to ensure the trust and safety of transactions.

The Ethereum blockchain, on the other hand, is a generic blockchain, on which you can write smart contracts and define the rules of the trust. In a way, if you can define trust in a code language, you can write smart contracts.

When not to use blockchain

Use blockchain technology as a tool for the development of specific scenarios. As we have seen in previous sections, blockchain technology can solve some problems where you want to replace an intermediary/middleman/TTP and do P2P value transfer transactions. This also includes problems where it is possible to replace existing trust with software-defined trust, using smart contracts.

You should not use Blockchain to replace your existing database. If your specification requirements can be solved with other software tools and techniques, then you should use those only. If there is no requirement for blockchain in your system architecture, avoid using it. For example, you do not need blockchain when building very high throughput, latency-critical transaction processing systems. However, when blockchain technology improves in terms of high throughput in the future, and you need some of the previously discussed properties of blockchain in your system, it would be a wise decision to use it.

Blockchain is slow

At the time of writing, the implementation of the Ethereum blockchain is slow and cannot handle high transaction throughput. There are some Ethereum upgrades scheduled to improve transaction throughput. Experiments and developments in improving transaction throughput for different blockchain implementations are still going on.

Let's compare the current situation of Ethereum with improvements planned for the future:

- As of now, Ethereum is like a single-threaded computer. It can process one transaction at a time.
- Sharding of blockchain would improve it and make it like a multithreaded computer. It would be able to process more transactions in parallel.
- The PoS consensus algorithm would help increase the reliability of the chain when a new block is generated in the blockchain.
- Layer 2 solutions, such as plasma and state channels, would enable off-chain transactions to process in near real time, and these transactions would be sent to the main chain periodically.

Considering the slowness and limited storage of Ethereum, it is not advisable to use Ethereum for pure database usage or for high processing power requirements. Storage on the blockchain is costly, so we should ensure that we know what we are trying to develop and whether it really requires the blockchain to be involved in its application architecture.

Blockchain depends on off-chain oracles

On their own, a blockchain is a separate network. It uses P2P networking over the internet to keep sharing transactions between nodes, but it does not have a way to get the data from outside (such as any data present on the internet, other than blockchain data) of the blockchain network, also known as **off-chain data**.

For example, the Ethereum blockchain cannot fetch any data from the internet on its own. There are some third-party developed Oracle services; by using them, you can fetch the data from the internet to the blockchain and do your transactions.

If your application is highly dependent upon third-party data available on the internet, then you should not use blockchain. If your off-chain data requirement is small enough, then you can use Oracle services and do your transactions on the Ethereum network. For example, when you just need currency version rates from the internet, you can use Oracle services.

When you are building a decentralized application on a public Ethereum blockchain and require off-chain data, you can use Oracle services such as Oraclize or ChainLink.

When you are building private applications using a private Ethereum blockchain and require off-chain data, you can write your own Oracle service application to read the data from off-chain. You can write your own Oracle to read some private data (such as data that's internal to the company) as well.

 You can set up your own private Ethereum blockchain on your private network. This private blockchain will not be connected to a public blockchain. You can customize your private blockchain according to your needs. However, applications deployed on a private blockchain would not be treated as purely decentralized applications.

Existing implementations of blockchain

There are many industries where blockchain technology can solve some problems. Let's take some examples and existing solutions that use blockchain technology to solve some bigger issues.

Cross-border payments

Cross-border payments are processed by multiple banks and take some time to process money transfers. Using blockchain, these banks can do cross-border payment transactions fast and almost instantly. Companies such as *Ripple* and *Stellar* are solving cross-border payments along with multiple banks. For example, in the year 2018 itself, HSBC bank moved $250 billion using DLT.

Decentralized cloud storage

You might be keeping your important files/pictures on centralized cloud storage such as Google Drive and DropBox. This centralized cloud storage is prone to attacks and these companies have full control over your files. By using decentralized data storage such as 3Box, you can store your public or private data on the blockchain.

Decentralized computing

If you have CPU resources available when you are not using your laptop or PC, then you can rent out your CPU power to a decentralized blockchain network. Someone who is in need of higher CPU usage would take your machine's CPU resources over the internet and, in return, pay you a small fee. For example, projects like Golem and iExec.

Introduction to Ethereum

Ethereum was proposed in 2013 by Vitalik Buterin, a cryptocurrency researcher and programmer. The first production release launched in 2015. Ethereum is an implementation of blockchain, and is an open source, public, and decentralized computing platform. Ethereum gained popularity because of its smart contract's features. Developers can program their smart contracts and execute them on the blockchain platform so that the full potential of the blockchain's different properties can be used.

Ethereum is for writing decentralized applications

As discussed before, Bitcoin uses blockchain technology for payments using bitcoins as a currency. Bitcoin also supports a scripting language, with which you can write small scripts. Bitcoin scripts have limited functionalities and it is difficult to write complex scripts on it. Its scripting language is also not Turing-complete (it does not support loops). Bitcoin transactions are slow; normally, it gets processed in 10 minutes or more.

With Ethereum, writing complex smart contracts and utilizing other properties of blockchain becomes easy. Any programmer who knows Java, JavaScript, and other programming languages can easily learn to code in the Solidity language. Solidity is a Turing-complete language. Transactions on Ethereum are processed much faster as compared to Bitcoin transactions. Using all of these potentials, you can write decentralized applications that would run on the Ethereum blockchain.

Ethereum architecture

Ethereum architecture consists of multiple entities that make its blockchain network. An Ethereum network has all the nodes connected to each other using the P2P network protocol; each node keeps the latest copy of the Ethereum blockchain ledger. A user can interact with the Ethereum network via the Ethereum client. The Ethereum client can be a desktop/mobile/web page:

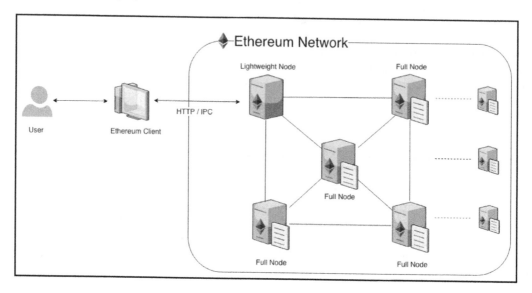

Ethereum high-level architecture

Let's discuss these entities.

P2P networks

In a P2P network, two or more computers are connected to share their resources without going to a centralized server or machine. Nodes connected in a P2P network can share or consume network resources.

Nodes

Nodes are devices in the blockchain network, which make the mesh of the blockchain network. Nodes can perform various types of tasks, such as being an active computer, cell phone, or disk drive as long as they are connected to the internet, and they participate in a blockchain network.

A node keeps the latest copy of the blockchain data (ledger) and, in some cases, processes the transactions as well. Miners put nodes on the network and willingly share their computing power or disk space. Nodes can be of two types—*full nodes* or *light nodes*.

Full nodes

A full node in a blockchain network maintains the complete data structure of the blockchain ledger and participates in the blockchain network by verifying every block and transaction using their consensus rules. Full nodes always keep the latest copy of the ledger.

Lightweight nodes

A light node does not maintain the full copy of the blockchain ledger. It verifies the transaction using a method called **Simple Payment Verification** (**SPV**). SPV allows a node to verify whether a transaction has been included in the blockchain or not. This is verified without the need to download the entire blockchain on a machine. A light node just downloads the header information of the blocks, hence lower storage space is required in running light nodes.

Miners

Owners of the nodes willingly contribute their computing resources or hard disk space to store and validate transactions; in return, they collect block reward and transaction fees from the transactions. The nodes that perform transaction execution and verification are called miners.

Blocks

The blockchain maintains a series of blocks. Each block is linked together with the last generated block. A block contains multiple transactions. At the time of writing this book, blocks are generated using a PoW consensus algorithm called **Ethash** in Ethereum. In future it is planned to change to PoS.

Miners share their processing power with the Ethereum blockchain network to process all pending transactions and generate a new block. Each block is generated in approximately 15 seconds. At the time of writing, a miner who generates the block is rewarded with 3 ETH in their wallet. It is proposed that, in the future, it will be reduced from 3 ETH to 2 ETH per block. It may change in future as well.

Each block has a defined gas limit: as of now, it is 8 million gas per block. This means that the total gas consumed by the transactions included in the block cannot exceed the 8 million gas limit per block.

A block contains a block header and transactions. A block can contain at least one transaction. A new block is added to the blockchain and linked to the previous block. The very first block in the blockchain is called the **genesis block**:

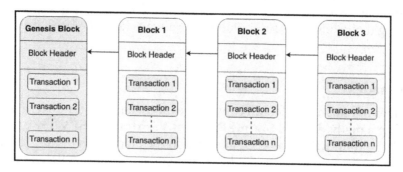

Chain of connected blocks

The block header contains metadata about the block and the previous block. It contains multiple fields, but it's only necessary to understand a few to have an idea of block generation:

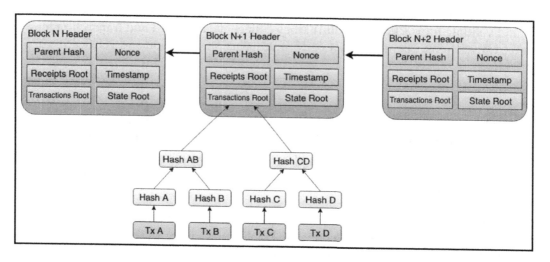

Block header and transactions root

The **Parent Hash** always points to the parent block of the newly generated block. This child-to-parent link goes on until the genesis block.

The **Nonce** is the number that is found by the miner to solve the cryptographic puzzle.

The **Receipts Root** is the Keccak 256-bit hash of the root node of the tree structure, which is populated with the receipts of each transaction.

The **Timestamp** is the time when the block is found and added to the blockchain.

The **Transactions Root** is the Keccak 256-bit hash of the root node of the tree structure, which is populated with each transaction.

The **State Root** is the Keccak 256-bit hash of the root node of the state tree, after all transactions are executed.

 You should know how hash functions work and what the properties of the hash functions are. Also, you should have a basic knowledge of asymmetric cryptography.

Ethereum Virtual Machine (EVM)

EVM is the runtime environment for smart contracts. Ethereum supports multiple scripting languages for writing smart contracts. These smart contracts are then compiled and converted into EVM bytecodes. Bytecodes are then executed by EVM to perform operations on smart contracts.

Ether currency

Ethereum has its own cryptocurrency called **ether** (symbol: **ETH**). Ether is a fungible coin, which means that a coin can be subdivided into smaller units. For example, 1 ETH can be subdivided into a maximum of 18 decimal places; the smallest value is called **wei**. Ether is a crypto fuel for the Ethereum network.

For any transaction you perform on the Ethereum network, some gas is consumed to execute that transaction, and the gas is to be paid in ether only. This transaction fee that you pay to process your transaction goes to the miners.

Ether is publicly traded on many centralized and decentralized exchanges. Some decentralized exchanges are also built upon the Ethereum blockchain itself using Solidity smart contracts.

Smallest unit – wei

The smallest unit of ether is denoted as wei. The conversion table for ether and wei is as follows:

Unit	Wei value	Wei	Ether value	Ether
Wei	1 wei	1	10^{-18} ether	0.000000000000000001
Kwei (KiloWei/babbage)	10^3 wei	1,000	10^{-15} ether	0.000000000000001
Mwei (MegaWei/lovelace)	10^6 wei	1,000,000	10^{-12} ether	0.000000000001
Gwei (GigaWei/shannon)	10^9 wei	1,000,000,000	10^{-9} ether	0.000000001
microether (szabo)	10^{12} wei	1,000,000,000,000	10^{-6} ether	0.000001
milliether (finney)	10^{15} wei	1,000,000,000,000,000	10^{-3} ether	0.001
ether	10^{18} wei	1,000,000,000,000,000,000	1 ether	1

The most commonly used terms are ether, gwei, and wei. These units are used in many places in the Ethereum ecosystem, such as when creating a new transaction, or when setting the gas price for a transaction. These units can also be used in the Solidity language, hence Solidity supports **wei**, **finney**, **szabo**, and **ether** units.

Gas

Gas is the fuel of the Ethereum blockchain network. Gas is always paid in ether. Gas is the transaction fee paid to miners for executing and adding your transaction into the Ethereum blockchain.

For every transaction, you have to specify the value of the *gas price* and *gas limit* you want to set. Based on these values, the network calculates how much of a fee you are willing to pay for a transaction.

Every bytecode operation in EVM has fixed gas units assigned. When that operation is executed by EVM, fixed quantity of gas is consumed by the transaction.

Gas limit

The gas limit is the maximum unit of gas your transaction may take, in order to be executed on the Ethereum blockchain. If your transaction takes less gas, then the extra gas that you have provided will be refunded back to your wallet. If your transaction consumes all the gas and requires more gas to be executed, your transaction will fail, and gas will be consumed as part of the transaction fees.

Gas price

The gas price is the price per gas unit you are willing to pay for executing your transaction. The unit of gas price is always defined in gwei. According to your gas price, miners decide to process your transaction. If the transaction has a higher gas price, there is a chance that your transaction will be processed early. If your gas price is low, then your transaction might be processed after a delay or when the blockchain is free to process low gas price transactions.

Formulas

The following are some formulas to calculate the *gas usage* and *transaction fee* per transaction:

$$TxFee_{Max} = Gas_{Limit} * GasPrice$$

$$TxFee_{Paid} = Gas_{Used} * GasPrice$$

$$TxFee_{Returned} = TxFee_{Max} - TxFee_{Paid}$$

$$TxFee_{Returned} = Gas_{Unused} * GasPrice$$

$$Gas_{Unused} = Gas_{Limit} - Gas_{Used}$$

$$BalanceRequiredAtTransactionInitiation = EtherValueToTransfer + TxFee_{Max}$$

Example

Let's take an example of a transaction where you want to transfer 0.1 ETH to your friend. The following are the values for the transaction fields you would have to fill while creating a transaction:

Field name	Value
From	0xff899af34214b0d777bcd3c230ef637b763f0b01
To	0xc4fe5518f0168da7bbafe375cd84d30f64cda491
Value	0.1 ether
Gas limit	30,000
Gas price	0.000000021 ether (21 gwei)

You can also refer to Chapter 4, *Learning MetaMask and Remix*, where we have demonstrated how to use MetaMask to initiate transactions.

Once the transaction is sent and confirmed on blockchain, you can see its status on block explorer (etherscan.io) as follows:

TxHash:	0x0a832532291aa34886d0a506f4062e25434b64c2b19a4fc0bd2ee2f4641cc186
Block Height:	9099873 (4 Block Confirmations)
TimeStamp:	30 secs ago (Oct-17-2018 12:38:00 AM +UTC)
From:	0xff899af34214b0d777bcd3c230ef637b763f0b01
To:	0xc4fe5518f0168da7bbafe375cd84d30f64cda491
Value:	0.1 Ether ($0.00)
Gas Limit:	30000
Gas Used By Transaction:	21000
Gas Price:	0.000000021 Ether (21 Gwei)
Actual Tx Cost/Fee:	0.000441 Ether ($0.000000)
Nonce & {Position}:	0	{0}

Screenshot from etherscan.io of the transaction

Let's calculate the values using these formulas:

- $TxFee_{Max} = 30,000 * 0.000000021 = 0.00063$ *ether*
- $TxFee_{Paid} = 21,000 * 0.000000021 = 0.000441$ *ether*
- $TxFee_{Returned} = 0.00063 - 0.000441 = 0.000189$ *ether*
- $Gas_{Unused} = 30,000 - 21,000 = 9,000$ *units of gas*
- $BalanceRequiredAtTransactionInitiation = 0.1 + 0.00063 = 0.10063$ *ether*

Let's go through the preceding calculations. When you initiate the preceding ether transfer transaction, it must have a balance of 0.10063 ether. If the balance in your wallet is lower than 0.10063 ether, then you will not be allowed to initiate transactions and this will be rejected by the client itself as it's an invalid transaction.

If you have sufficient balance and the transaction is initiated, then, along with your transaction, the *BalanceRequiredAtTransactionInitiation* amount will be locked. So, in this example, 0.10063 ether is locked.

Now, your transaction is processed, 21,000 gas has been consumed in order to execute the transaction on the EVM, and the transaction executed successfully. As it consumed only 21,000 gas, it refunds 9,000 units of gas. This means 0.000189 ether will be refunded back to your wallet and you have paid 0.000441 ether for your transaction processing.

You can remember this Gas_{Limit} and Gas_{Price} with a car and fuel example. Let's say you want to go from city A to city B and the distance between these two cities is 500 km. Your car fuel average is 20 km/liter ($Gas_{Required}$). There is no fuel station on the way when you travel, so you would have to get it filled before you start. You need $500/20 = 25$ liters ($TxFee_{Required}$) of fuel to reach city B. If you filled your car fuel tank with 30 liters ($TxFee_{Max}$), then you would successfully reach city B and you will still have 5 liters (Gas_{Unused}) left with you that you can use later on. However, if you had filled it with 20 liters when you started the journey, then you would have stopped at 400 km and you couldn't have done anything. This would be called a failed transaction and your gas/fuel is consumed so there's no gas refund.

An ether transfer requires a fixed 21,000 gas limit per transaction. If you also include data along with the ether value, it may require a greater gas limit to execute the transaction.

Ethereum accounts

On the Ethereum platform, you can have two types of accounts:

- **Externally Owned Accounts (EOA)**
- Contract accounts

Both of these accounts can hold balances of ETH; however, there are some differences between these accounts, which we will cover in the following sections. Both account addresses are in hexadecimal, which starts with the 0x prefix. The Ethereum public address of an EOA and a contract is always a **20 bytes (160 bits)** address represented in hexadecimal form:

Ethereum account/contract public address format

You need these accounts to perform transactions on the Ethereum network.

Externally owned accounts

An Ethereum wallet can hold one or more EOA. Wallets can be created by anyone via wallet creation websites, browser plugins such as MetaMask, or mobile apps. These wallet creation services generate a new private key using a specific account creation algorithm. Using this private key, the public key is derived. The combination of a private key and a public key is called an **account** or EOA. A public key can be shared with anyone to receive ETH/ERC20 **tokens**. However, the private key must be kept secret by the user. For the creation of wallets using MetaMask, you can refer to Chapter 4, *Learning MetaMask and Remix*.

An EOA is controlled by its private key to initiate any transaction with it. An EOA owner can send ETH/ERC20 tokens to other EOA or contract addresses. They can also initiate a transaction on a deployed contract with EOA:

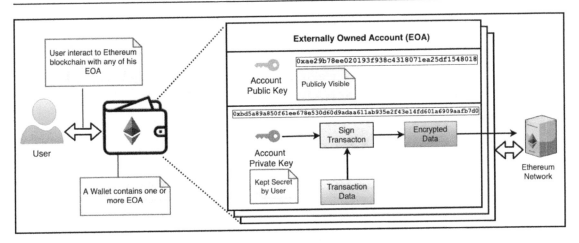

User interacting with Ethereum network using the EOA present in their wallet

As shown in the preceding diagram, the user can interact with the Ethereum network via one of the EOA accounts present in his wallet. A user holds a wallet that can have one or more EOA accounts. Each EOA account has a private key, which the user must use so they can sign the transaction data and send it to the Ethereum network to be executed. The user can also check their account balance from the Ethereum network using their public key.

 New wallets are created off-chain using wallet generation algorithms used by MetaMask or other available websites. They are not created on-chain; however, smart contracts and their addresses are created on-chain when a contract is deployed on the blockchain.

Contract accounts

This type of account is also known as a **contract**. Contracts only have a public address, and their code logic controls the flow of the funds and other states of the contract. Contracts do not have any private keys associated. You need an EOA account to create a new contract or to interact with an existing contract. Contract accounts have a Solidity code that defines their behavior.

Any Solidity code you deploy creates new contract accounts or contracts:

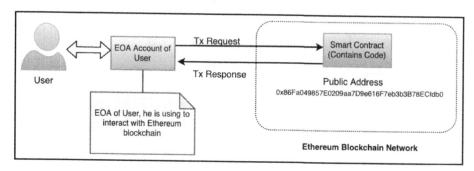

A user accessing a contract account via their EOA account

As shown in the preceding diagram, the user has their own EOA account. Using this EOA account, the user can interact with any deployed contract accounts (smart contracts) present on the Ethereum blockchain. The user can interact by initiating transactions. Also, with an EOA account, the user can deploy a new contract.

Contract accounts can only be created on-chain and their contract bytecode is included in the block when they are deployed successfully. Contract accounts are not created off-chain, just like EOA accounts.

The difference between an EOA and a contract

There are some differences between an EOA and a contract account. The following table distinguishes between an EOA and a contract:

	EOA (account)	Contract account (smart contract)
Creation	A new account can be created at any time, using a wallet generation utility such as MetaMask or with `https://www.myetherwallet.com/`.	A new contract can only be created from an EOA or from an existing deployed contract.
Public address	An EOA account's public address is derived from its private key.	A contract's public address is created by the combination of a *public address of creating account + nonce*.
Private keys	An EOA can only be controlled using its private key.	A contract which do not has a private key can be controlled using other EOA or contract. They respond to transactions initiated from an EOA.

Control	Whoever owns a private key can control the funds of an EOA.	Control is defined by the code stored with it. An EOA or another contract may have control over a deployed contract.
Code	An EOA does not have any associated Solidity code.	A contract always has Solidity code associated with it, which is stored together with it.
Access	Using the private key of an EOA, we can make transactions via different wallet software such as MetaMask and Ethereum wallet.	Using **Abstract Binary Interfaces** (**ABIs**) of a contract, an EOA can access the functions of the contract via Remix, Truffle, and others tools.
Transaction initiation	Using the private key, one can directly access the funds of EOA and initiate any transactions. So, direct access to funds is allowed.	A contract cannot initiate a transaction on its own. An EOA need to initiate the transaction first, then only contracts can initiate other transactions.

As we have discussed the basic differences between an EOA and a contract, let's discuss Ethereum transactions.

Ethereum transaction

A transaction on an Ethereum blockchain is required to transfer ETH from EOA or for smart contracts to change state, perform an action, or read a state of a contract. A transaction can only be initiated from an EOA and it can further initiate other internal transactions.

For example, an EOA wallet initiates a transaction to call a function of contract A, and this contract calls another function of contract B. In this case, a transaction is initiated from an EOA to contract A and then an internal transaction is initiated from contract A to contract B.

Transaction fields

A user first fills the required fields for creating a new transaction. Then, they sign the transaction with their EOA's private key. This signed transaction data is then broadcast to the Ethereum blockchain network via any Ethereum client. The Ethereum client also first verifies that the transaction is valid and signed using the EOA's private key only, the address of which is in the *from* field.

From

From is always the public address of the EOA that initiated the transaction. In Ethereum, only an EOA account can initiate a transaction. Contract accounts cannot initiate a transaction on their own. The transactions further initiated by contracts are called internal transactions (on `etherscan.io`) in general terms; however, as per the Ethereum's yellow paper, these transactions are called message calls.

To

To is always the public address of an EOA or a contract. When *To* is an EOA address, then this transaction is initiated to send some ETH. When *To* is a contract address, then this transaction is initiated to call a method of a contract or to send ETH to contract..

Value

A transaction can have either zero or nonzero for the value field. Normally, the value field is nonzero when sending ETH to an EOA. However, when initiating a transaction to a contract, the value field either has a zero or nonzero value based on the contract function.

Gas limit

The *gas limit* is the maximum limit of gas units that will be required by the EVM to execute the transaction. We discussed this in detail earlier in this chapter.

Gas price

The *gas price* is the cost per gas unit you are willing to pay. We discussed this in detail earlier in this chapter.

Nonce

The *nonce* field in a transaction always starts with 0 (zero) for an EOA that initiates the transaction. The nonce keeps increasing by one for every transaction an EOA initiates. The nonce also specifies the execution order of the transactions initiated from an EOA. When multiple transactions initiated from an EOA are in the pending state, then the EVM executes a transaction that has a lower nonce value for that EOA.

Data

Data represents the hexadecimal data that you want to transfer along with a transaction. This data field is required to call a method of a contract.

When transferring ETH from one EOA to another EOA, the data field is empty, which means the value of the data field will be `0x`. However, when we want to execute a method of a contract, the data field will hold the hexadecimal value of the method and its arguments:

```
0xa9059cbb00000000000000000000000055db3f5a1543e1717661a77b0f65b4e0c6a6
922600000000000000000000000000000000000000000000000a2a15d09519be00000
```

Transaction data in hexadecimal from etherscan.io

The preceding hexadecimal transaction data, once decoded, represents the `transfer()` method call (called on an ERC20 contract) with the `_to` and `_value` parameters passed in. As you can see, `0xa9059cbb` (four bytes data) is the hex code for the `transfer()` method, and the following method parameters are passed in:

```
Function: transfer(address _to, uint256 _value) ***

MethodID: 0xa9059cbb
[0]:    00000000000000000000000055db3f5a1543e1717661a77b0f65b4e0c6a69226
[1]:    00000000000000000000000000000000000000000000000a2a15d09519be00000
```

Hexadecimal transaction data decoded on etherscan.io

Once the previous transaction data is decoded on the `etherscan.io` website, it shows the first four bytes of the function signature and the parameters passed to the function.

Transaction hash

A user doesn't have to set the transaction hash field. A transaction hash is assigned to every transaction when it contains valid transaction fields and is broadcast to the Ethereum blockchain. The transaction ID is the SHA-256 hash of the signed transaction data. Using this transaction ID, we can track the status of the transaction and can cancel or drop the transaction before it gets executed.

Transaction status

The transaction status lets the user know the status of a transaction. Transaction change or dropping a transaction is allowed in different states. A transaction can be in any of the following three statuses:

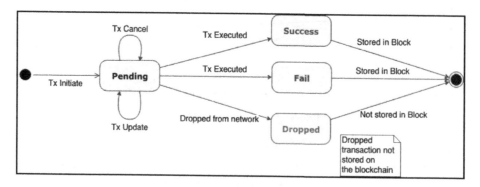

Transaction status diagram

As shown in the preceding diagram, transactions are always initiated by an EOA. Let's go through each of the statuses.

Pending status

When a transaction is submitted to the Ethereum network, it first goes into the pending status, waiting to be executed by the nodes. A transaction can be in the pending state for a longer duration if the gas price is set very low in the transaction and the nodes are busy processing other higher gas price transactions.

During the pending state, the transaction initiator is allowed to change the transaction fields at any time. They can do so by sending another transaction with the same nonce.

Success status

A transaction that is executed successfully without any failure is called a **successful transaction**. A successful transaction means the intended operation is executed successfully by the Ethereum blockchain and verified by other nodes present in the Ethereum blockchain network:

TxHash:	0x3483462e641001685f0a2ac5a29d400db04c8a2515e0f0d531fd5970e28892b5
TxReceipt Status:	Success
Block Height:	6555795 (1 Block Confirmation)
TimeStamp:	21 secs ago (Oct-21-2018 10:04:46 AM +UTC)
From:	0xff62c51fd4c72e5ad5b00f7a7eccecb30c5ce172
To:	0x53d66aca4a231e6568d16397c7b117d71d10c506

A successful transaction on etherscan.io

Fail status

A transaction is executed and, during the execution, it may fail due to an error in the contract method execution. Transaction failure also depends on the contract definition. Transaction execution failure could also occur in case of a low gas limit, and transactions will fail with the OutOfGasException error:

TxHash:	0xd4a0dc1c7d6129408b6d49101be32df2fe75297d6057f891cdad06f311e19c8a
TxReceipt Status:	Fail
Block Height:	6555182 (58 Block Confirmations)
TimeStamp:	11 mins ago (Oct-21-2018 07:49:01 AM +UTC)
From:	0x86fb126ed3c5b7be5b531d7a0bef68f8365d1ee7
To:	Contract 0x86fa049857e0209aa7d9e616f7eb3b3b78ecfdb0 (EOSTokenContract) ⚠
	└ Warning! Error Encounter during Contract Execution [Reverted] ☹
	└ ⚠ ERC-20 Token Transfer Error (Unable to locate Corresponding Transfer Event Logs), Check with Sender. ⓘ

Failed transaction on etherscan.io

Dropped status

A transaction is sometimes dropped from the Ethereum network for several reasons:

- If the transaction was in the pending state for a very long duration
- It has a very low gas price as compared to other pending transactions
- The maximum number of pending transactions the Ethereum blockchain can handle are in the pending transaction pool

Once a transaction is dropped, all the gas and ETH is returned to the originating wallet:

Transaction Information {Dropped/Replaced}	
TxHash:	0x1943228853a9ab55affde84e3e7b99e5bf9ce502f55f94993686f011a508c7c2
Replaced by TxHash:	0x83873da30ddab1a5d1a468cb67e737a53551eebc3400652508bb9311d2ae3bb9
Block Height:	- n/a -
Time FirstSeen:	Jan-28-2018 09:01:22 AM
From:	0x876eabf441b2ee5b5b0554fd502a8e0600950cfa (Bitfinex_Wallet4)
To:	0x1928ff42b3e0eb0e3e4aa80f10fdef704c74ac8f

Transaction dropped on etherscan.io

The dropped transaction is not included in any block on the blockchain, as if it didn't happen or wasn't initiated. There is no trace of those transactions on the blockchain; you can only see these dropped transactions on `etherscan.io` (block explorer) for some time. Later, they may be removed from `etherscan.io` as they are not part of the blockchain. However, successful and failed transactions are included in the block of the Ethereum blockchain and their status can be checked from any blockchain client.

Transaction operations

Two types of operations are allowed on an already initiated transaction. To do either of these operations, a transaction must be in the pending status.

Replace/update

If a user wants to change some transaction fields they have initiated, they can do so by changing the values and keeping the same nonce in the new transaction. A replace transaction is mostly used when a transaction is in the pending status for a longer duration because of the low gas price, and just increasing the gas price of the same transaction could increase the likelihood of it getting executed by the Ethereum network.

Cancel

A transaction can be canceled by the same EOA that initiated the same transaction. If a user feels that they have initiated a wrong transaction accidentally, they can cancel that transaction by initiating another transaction having the same nonce and changing the *To* fields to their own EOA address. This new transaction must have a higher gas price. A higher gas price is set to ensure that the new transaction will be given priority over the older transaction and can cancel the transaction as soon as possible. This type of transaction is called a **self transaction**.

Testnets

There are multiple testnets available where you can deploy and test your contract's functionality. These testnets are the replica of the mainnet (production) Ethereum blockchain, so these testnets also behave the same. If your contracts are working on testnets, then they will work perfectly on the mainnet as well.

However, to deploy and test your contracts on these testnets, you need to have some testnet ethers for that testnet. You can get those testnet ethers from a faucet for free from the following locations:

Testnet	Faucet link	Block explorer
Kovan	https://faucet.kovan.network/	https://kovan.etherscan.io/
Rinkeby	https://faucet.rinkeby.io/	https://rinkeby.etherscan.io/
Ropsten	https://faucet.metamask.io/	https://ropsten.etherscan.io/

If you want to test your contracts locally, then you can either install and set up your private testnet or you can use the Ganache tool.

Smart contracts

Nick Szabo coined the term **smart contract** in 1994. Smart contracts are self-executing contracts. They eliminate intermediaries and allow direct P2P transactions. A smart contract has the power to automatically enforce some obligations. One of the best examples of this is a vending machine: when you need an item from a vending machine, you pay into it and it gives you that item without needing any middleman.

Smart contracts are very different from traditional software programs. Smart contract code is immutable once deployed on the blockchain. Because of this immutable property, we have to think twice before deploying it; otherwise, it could end up with critical flaws in production. A different mindset is required for writing smart contracts. Some smart contracts present in decentralized applications handle the flow of ETH or other ERC20 tokens, which have real monetary value. If something goes wrong in the production, any attacker can steal ETH or ERC20 tokens from your smart contracts.

Immutable code

Once you deploy your smart contract on Ethereum blockchain, either on the testnet or the mainnet, you cannot modify the code of the smart contract files that you have deployed. All code that is deployed becomes immutable. Not even an attacker/hacker can change the deployed code.

If your code is deployed on the testnet for testing purposes, then you can fix any issues in your smart contract files present on your local machine, and redeploy your code on the testnet again to check whether the functionality is working or not. But if you have deployed your contract on the mainnet (production) and other people, entities, or contracts started using that contract code, then you cannot change that contract code. If those entities are still not using that freshly deployed code, then you still have a chance to fix any issues/errors and redeploy. There is another way to create an upgradable contract in Solidity; for more details, you can refer to Chapter 11, *Upgradable Contracts Using ZeppelinOS*.

You cannot change the code of smart contracts, but you can change the different states/variables of the smart contracts—those are handled with smart contract code. Based on the different states/variables, your smart contract code behavior changes.

Irreversible transactions

Once the smart contracts are deployed on Ethereum network, their code becomes immutable. No one can change the code of the deployed smart contract. It's the same as when some transaction on the Ethereum network is done, and there is no way to reverse that transaction. For example, if person A sends 1 ETH to person B and the transaction is successfully executed, then there is no way for person A to reverse that transaction and to get that 1 ETH sent back to their wallet.

For smart contracts as well, if you have done a particular transaction on a smart contract that has changed the state of the contract, then there is no way to reverse that transaction. There might be some other functions available in the smart contract to get the same state again, but, in terms of transactions, it is not possible to reverse the transaction that was executed.

Think twice before deploying

I am serious—think twice before deploying smart contracts on production!

Before deploying smart contracts on the mainnet (production), you must ensure that all of the unit tests and other test cases that you have for the smart contract are passing. Your code is reviewed by some other developer to ensure that the logic of the contract is correct and is going to behave as per your expectations in production. Ensure that the code coverage reports are reviewed, and the maximum possible coverage of the code is ensured.

You might be wondering why I am mentioning these normal things—as a developer, you would think that these are the processes we follow on a daily basis during software development, so what's new and special about this?

To date, as a software developer, you have been writing software/applications and following best practices to ensure that no errors/bugs are in the system. But, at the back of your mind, you also think that if something goes wrong with the script/code, we can apply a patch and re-deploy the software/application again, with no issues. Since the beginning of software development, this has been the practice.

Now, as a software developer, you need to understand that smart contracts are not like your existing software/applications, where you can apply a patch and change the code at any time. As smart contract code becomes immutable once deployed, only states/variables can be changed via allowed methods that are present in smart contracts. As a developer, you need to mentally prepare yourself for the fact that the smart contracts you are writing cannot *ever* be changed once deployed on the Ethereum blockchain. If you have experience of writing device drivers for hardware, then you can understand this easily as device drivers have limited changeability/upgradability.

If you are planning to upgrade smart contracts in the future, then you need to be prepared for it when deploying the first version of the contracts. For example, you should be able to stop an existing version of a smart contract and move on to the new upgraded version. You also need to think about how existing users will migrate to a new upgraded version of contracts.

There are some design patterns available to provide upgradability to smart contracts. We will cover those in `Chapter 11`, *Upgradable Contracts Using ZeppelinOS*.

The preceding summary is not intended to threaten you about how smart contracts could go wrong, but to inform you about the consequences it could cause if the code is not well tested and reviewed with multiple iterations.

Limited storage

When your smart contract stores some variables/states, it has to be stored on the blockchain's ledger. The storage space of smart contracts is limited and costly. As the gas limit per block is limited, so is the execution of a transaction. The Ethereum blockchain should not be used as data storage or a replacement database. The blockchain is a kind of decentralized database; however, at the moment, the Ethereum blockchain cannot handle the processing of loads of data in a transaction. Blockchain technology is still evolving; in the future, it might be possible to even store large amounts of data and processes in a single transaction.

If you want to store big files or data on decentralized blockchains, you should look for Swarm, IPFS, StorJ, or Filecoin, according to your needs.

Every transaction consumes gas in ether

The gas consumption that we are going to talk about is with respect to the Ethereum blockchain. We are not going to talk about how other types of blockchain treat their smart contract executions.

On the Ethereum blockchain, every transaction that sends ether consumes gas; in the same way, to execute a smart contract's logic, the transaction initiator has to pay gas in ether. As a developer, you must ensure that the logic of your smart contract is not complex and consumes optimal gas. You might need to re-design your logic to ensure that gas consumption is not very high.

Playing with ether or tokens

As we saw previously, the main currency of the public Ethereum blockchain is ETH. There are some other currencies also present on the Ethereum blockchain itself; these follow ERC20 standards and are called tokens. ETH and some of the ERC20 tokens are being traded on some cryptocurrency exchanges.

You can write some smart contracts that involve ETH and ERC20 tokens in your contracts; for example, if you are writing a decentralized exchange on the Ethereum blockchain. To name a few, some existing decentralized exchanges include EtherDelta, IDEX, and KyberNetwork.

When using ETH/ERC20 tokens in your smart contract, you need to ensure the safety of funds. The following are some examples that have led to multimillion dollar hacking of smart contracts on the Ethereum blockchain itself. One of the famous attacks was the *DAO Smart Contract* hack, in which attackers were able to spawn new child contracts and were able to steal ETH from the main contract. Another famous attack was the *Parity multi-signature wallet* hack, in which attackers were able to steal ETH from all of the contracts that were using the Solidity code of the Parity multisignature wallet. To prevent this attack, white hat hackers have to hack all other existing Parity multi-signature wallets, as there was no other solution and no way to change the contract code.

Not all smart contracts deal with ETH/ERC20 tokens, but if they are dealing with ETH/ERC20 tokens, then extra care must be taken before deployment.

Summary

Blockchain has some unique properties that make it powerful. It uses all the benefits of decentralized systems. Blockchains can replace the TTP (the middleman) and provide the required trust. The irreversibility of transactions makes blockchain secure and attack-proof.

Ethereum is one of the implementations of blockchain. Bitcoin blockchain has its scripting language, but it is not Turing-complete. Ethereum supports the Turing-complete Solidity language, which we can use to write smart contracts. At the moment, Ethereum transaction processing is slow. In the future, it is going to have some upgrades, which could improve the transaction processing speed of the overall Ethereum blockchain network.

Blockchain makes smart contracts immutable, which opens up a new way of programming in the new paradigm. Smart contracts can cause problems if deployed on production without due diligence and testing.

In the next chapter, we will learn about the most popular smart contract language, Solidity. We will go through contract structures, data types, global variables, and more.

Questions

1. Why do we need blockchain?
2. Can blockchain replace existing database technologies?
3. Can we use blockchain to solve every problem that can we do with coding?
4. Why do we need to remove intermediaries?
5. What are the limitations of Ethereum?
6. Why are the gas limit and gas price required in a transaction?
7. What are the benefits of having immutable code and irreversible transactions?
8. Why are smart contracts different from traditional programs?
9. Why is the testing of smart contracts a must?

Getting Started with Solidity 2

In the last chapter, we learned about blockchain, smart contracts, and Ethereum. For writing smart contracts, there are multiple languages supported by Ethereum, including Solidity, Vyper, and **Low-level Lisp-like Language (LLL)**.

In this chapter, we will deep dive into Ethereum's most popular contract-oriented language, called Solidity. We will cover Solidity's source code file structure and the different data types supported. We will also understand the different units, global variables, and functions it supports and where these should be used.

We will cover only the parts of Solidity language that are essential for Solidity developers. We will not cover the advance concepts of Solidity language.

The following topics will be covered in this chapter:

- The structure of a Solidity source code file
- The structure of a Solidity contract
- The different data types and units supported by Solidity
- The global variables and functions available

Introduction to the Solidity language

Solidity is high-level language oriented for smart contracts. It is a Turing-complete language, influenced by C++, Python, and JavaScript.

Solidity is a statically typed language that supports inheritance, libraries, and complex user-defined types. Solidity is a bit different from the other languages; for example, Solidity does not have *null* values. Variables are initialized with their default values. Solidity also does not have any exception-handling mechanism to propagate exceptions.

Solidity support different data types including `uint`, `int`, `address`, and many more. As developers, we must know that Solidity does not have boundary checks for the `uint` and `int` data types. This makes these data types vulnerable to overflow or underflow.

At the time of writing this book, a major release of Solidity, version 0.5.0, has come out. This release is not fully backward compatible with the previous release of version 0.4.25. This means if you have a contract compiled with version 0.4.25, it may not compile with 0.5.0 and could produce some compilation errors. However, if your contracts are deployed on the Ethereum blockchain compiled with 0.4.25 or previous versions, you can still interact and call functions on those contracts from a version 0.5.0 contract via interfaces.

Another thing to notice here is that, as the new version 0.5.0 has been released, it doesn't mean that all of your smart contracts must be compiled and developed using version 0.5.0 only. You can still use old compiler versions to develop your contracts. The only benefit of using version 0.5.0 is that you will benefit from some advanced features, such as security features that developers can use, including the newly added `address payable` data type, which increases the safety of the ether funds in a contract.

There are a few improvements added to Solidity version 0.5.0. In this book, we will mostly focus on Solidity version 0.4.25, but where some new features or changes introduced in the latest version 0.5.0 are relevant to the topics we'll discuss, then we will also briefly examine these too.

We will discuss and explain some examples of code during this chapter. If the code itself is not explicitly using any `pragma solidity` directive, it is assumed that the code is using Solidity version 0.4.25. However, if the code syntax requires some other Solidity version, such as version 0.5.0, then it will be explicitly specified in the first line of the code.

We will not cover all parts of the Solidity language. We will only go over those areas required to write basic smart contract functionality. There are many advanced features supported by Solidity, such as assembly. We assume that some very advanced features will be self-learned by the reader when required.

The layout of a Solidity source file

A Solidity source file can contain any number of contracts/library definitions in a single file. It is recommended that the developer should maintain the different source files for each contract/library definition to have the readability and code maintainability. The Solidity contract/library file extension is `.sol`, for example, `SampleContract.sol`.

The basic layout of the Solidity source file contains following in order:

- The Solidity version to use
- Experimental `pragma` if any
- Import statements to `import` other contracts/libraries
- Contract definition covered in the *Structure of Contract* section in this chapter

Solidity version with pragma

Each Solidity contract/library source file should have the version of Solidity compiler specified. The version of the compiler should be specified first thing in the source file. The following is an example of how to specify the Solidity version:

```
pragma solidity ^0.4.4;
```

If a source file is of this version, it means the file will compile with 0.4.4 and higher versions in the 0.4.x branch, up to version 0.5.0 where 0.5.0 is not included.

It is recommended that the Solidity version should be locked and specified without the ^ (caret) sign, as follows:

```
pragma solidity 0.4.4;
```

This ensures that the specific compiler version will always be used, using which code is tested most. It also prevents the code from exhibiting any unintended or unexpected behaviors (such as bugs introduced with new compiler changes).

Importing other source files

Solidity supports an `import` statement to import other source files. The following is the syntax for importing:

```
import "path/to/file/filename.sol";
```

In the preceding example, the absolute path of the file should be specified. This absolute path of a contract must be inside the project's `contracts` folder. The `contracts` folder will contain all Solidity contracts in a Truffle project. You can refer `Chapter 5`, *Using Ganache and the Truffle Framework*.

One important point to note here is that the `import` statements are for your development environment only. For example, the Truffle framework does understand the packaging of the contracts and `import` statements that makes it easy to maintain and work on multiple `.sol` files. However, when you deploy your contracts on the Ethereum blockchain, `import` statements must be replaced with the actual content of the `.sol` file that you imported. The Ethereum blockchain takes the flattened files of the smart contract and deploys it. The framework, such as Truffle or the Remix IDE, converts all of your `import` statements and makes a single flattened contract when you deploy your contract on the blockchain.

Structure of a contract

Just like in an object-oriented language such as Java, where we have classes, similarly we have contracts in Solidity languages. Each Solidity contract can contain declarations of one or many state variables, constants, a constructor function, enumerations types, function modifiers, events, functions, and struct types. Furthermore, a contract can inherit from other contracts as well.

You can also define special kinds of contracts called libraries and interfaces. Libraries are deployed only once along with your contract deployment. Interfaces cannot have function definitions, variables, struct types, or enums. We have a separate chapter where we will talk about libraries and interfaces in detail.

Normally, the structure of these constructs is maintained in the following order. Although there is no hard rule for it, it would be good to maintain this order so that other developers can also understand the code at a glance:

1. Global state variables of contract
2. Enum types
3. Struct types
4. Function modifier definitions
5. Events definitions
6. Functions

We will have a brief overview of the state variables, functions, events, and struct types that create the contract structure. We will discuss these topics in detail in the further sections of this chapter.

Declaring state variables

The state variables defined in contract are values that are permanently stored in contract storage over the Ethereum blockchain. The variables defined outside of the method/function body are accessible to the whole contract. Based on the visibility defined for the variable, it may or may not be further accessible from other inherited contracts.

Defining a storeUint variable of the uint type is done as shown in the following code:

```
contract VariableStorage {
    uint storeUint; //uint256 storage variable
    //...
}
```

The storeUint variable in the preceding code block will be accessible from any function/modifier defined in the VariableStorage contract.

Writing function definitions

The Solidity functions are executable units of code within a contract. These are the only access points for the users and the different entities that will interact with a contract. Functions may use function modifiers; they also have a visibility to define the access rights. There can be many functions within a contract.

A deposit function is defined as shown here. Your contract can receive ether along with a method call using the payable keyword, as shown in the following code:

```
contract AcceptEther {
    function deposit() public payable { //function example to accept ETH
        //...
    }
}
```

Creating a custom modifier using function modifiers

The modifiers are used to check the pre- and post-conditions before calling a method. Argument validation and correctness are checked with the modifiers. The function modifiers are also used to maintain the access and role-based control on the contract.

In the following example code, onlyOwner is an access modifier function used in the withdraw function. The definition says that the withdraw function can only be allowed to call from the owner address:

```
contract AccountContract {
    address public owner;

    /*
     * Modifier onlyOwner definition.
     */
    modifier onlyOwner() {
        require(msg.sender == owner, "not owner");
        _;   //Rest of the function body execution
    }

    /*
     * The deployer of the contract would become owner of contract
     */
    constructor() public {
        owner = msg.sender;
    }

    /*
     * Modifier onlyOwner used, only allow owner to
     * call withdraw function
     */
    function withdraw() public onlyOwner {
        msg.sender.transfer(address(this).balance);
    }

    //...
}
```

In the preceding code, in the onlyOwner modifier, we have used require() function to check that the sender is the current owner of the contract; otherwise, the transaction will be failed with a not owner error message.

Using events for logging and callback

Solidity contracts do not have a way to generate logs when they are executed. Events are used to log values of the variables that can be listened as a callback by any JavaScript client. When events are emitted, its data is stored in the transaction's log.

Contracts emit event logs to let the other parties know that some action has been performed on the contract. Events also provide a way to maintain action history. Client applications can listen for the events as a callback and can take appropriate action when required. Client applications can also search for a specific event from the event history of the contract.

Deposited is an event in the following code, which is emitted when the deposit function is called:

```
contract AcceptEtherWithLog {
    /*
     * Event declaration
     * @param _who Address of the account who deposited ETH.
     * Parameter indexed, to allow it to filter events on client side.
     * @param _amount Amount of Wei deposited to contract.
     */
    event Deposited(address indexed _who, uint256 _amount);

    function deposit() public payable { //function example to accept ETH
        //...
        //Emits the Deposited event in Logs of transaction data
        emit Deposited(msg.sender, msg.value);
    }
}
```

The Deposited event logs two parameters, _who and _amount. As you can see, _who is an indexed argument, and Ethereum creates the topic index for _who values. This enables client applications to quickly search for specific events with a specific _who address.

Custom data types with struct

Just like other languages such as C++, Solidity also supports defining structs. Structs are custom-defined data types that can group multiple variables in a single struct variable to form a new type.

By default, all `struct` values are initialized using their default values. For example, `uint` values will be initialized with `0`; `bool` values will be initialized with `false`; and address values will be initialized with `0x0`, as follows:

```
contract LoanStruct {
    //Definition of struct
    struct LoanData {
        address borrower;
        address lender;
        uint256 loanAmount;
    }
}
```

 `0x0` is the short form of a specific Ethereum address, **0x0000000000000000000000000000000000000000**. This address is also used for logging purposes, especially when minting new tokens. We will have some examples of these in `Chapter 9`, *Deep Dive Into OpenZeppelin Library*, dedicated to OpenZeppelin libraries.

Custom types for constants with enum

Like other languages such as C++ and Java, Solidity supports enumerations. Enums can be used to create a custom data type with a fixed set of *constant values*. Enums will have integer values assigned, starting from `0` and ascending to the number of states defined in `enum`. Enums in Solidity adopts to `uintX` data types starting from `uint8` up to `uint256`. According to the number of values present in `enum`, compiler chooses the data type to be used. For example, when having 1 - 255 number of values defined in an `enum` would use `uint8`, however, it will use `uint16` when a few more than 255 values are used.

As shown in the example code here, by default, the `status` variable will be set to `0`:

```
contract LoanStruct {

    //Enum for LoanStatus
    enum LoanStatus {Pending, Created, Funded, Finished, Defaulted}

    //Definition of struct
    struct LoanData {
        address borrower;
        address lender;
        uint256 loanAmount;
        LoanStatus status; //LoanStatus stored.
    }
}
```

Solidity data types

The Solidity language is statically typed. When writing a contract or library definition, all of the state variables and local variables must have the data type specified along with its declaration. There are many data types supported by Solidity. You can create your own custom type by using the data types supported by Solidity; for example, you can define your complex `struct` variable.

Solidity also has a support for the `var` keyword, used for dynamic data type. However, since Solidity version 0.4.20, the `var` keyword has been deprecated because of security concerns.

If we look at the Java language, it has primitive data types and reference data types. For reference data types, especially for instantiable class, the default value is null while initialization. However, in Solidity, there is no concept of null values. Instead of null, the default value of each data type is defined. If there is no initial value assigned to a variable, it will be initialized with its default value. Refer to the following table for the default values of each data type:

Solidity type	Default value
`uint8` - `uint256`	0 (zero)
`int8` - `int256`	0 (zero)
`bool`	`false`
`string`	An empty string
`byte`	In integer form, 0; in hex form, `0x00`
`bytes1` - `bytes32`	In integer form, 0; in hex form, `0x<NumberOfZeros>` (*NumberOfZeros = NumberOfBytes * 2*; for example, for `bytes2`, the default value is `0x0000`)
`bytes`	Empty bytes prefixed with `0x`
`address`	`0x0000000000000000000000000000000000000000`

Defining default values of Solidity types

As for `int` and `uint`, we have different sizes available, just as for `bytes`, we have `bytes1` to `bytes32` available. However, don't get confused here if you are using `bytes` as it's an array of `byte[]`, and `bytes1` to `bytes32` are specific size of bytes. Hence, `bytes` and `bytes1...bytes32` are both different.

Before any function execution, the first thing you should ensure is that the arguments passed into the function of any data type are either expected, allowed, or under the bound you have defined. To validate these arguments, you can use the `require` statement.

Solidity's data types are further divided into the following:

- **Value types**: The data types whose values are copied when passed as method arguments. These are `int`, `uint`, `bool`, and `address`.
- **Reference types**: The data types whose references are copied when passed as method arguments. These are `array` and `struct`.

Understanding Solidity value types

The `int`, `uint`, `bool`, and `address` data types are also called **value types**. The variables of these types will always be passed by value. This means that their values are always copied when they are used as the function arguments or assigned to another variable.

Some `int` and `uint` value types in Solidity support various sizes to reduce gas consumption and storage space. You should choose the appropriate value type based on your data requirements. The `bool` and `address` value types do not have different sizes, as their size is fixed. The size of the value types is as follows:

Value type	Different sizes	When size is undefined
int	int8, int16, int24 .... int256 (a step of 8)	int uses int256
uint	uint8, uint16, uint24 .... uint256 (a step of 8)	uint uses uint256

The different sizes of value types supported

Integer value type

Solidity supports signed integer (`int`) and unsigned integer (`uint`) data types of various sizes. Based on the size of your data, you can choose from 8- to 256-bit (in steps of 8) data types. For example, for `int`, you can have `int8`, `int16`, `int24`, and so on, in increments of 8 up to `int256`. In the same way, you can have `uint` from `uint8`, to `uint16`, `uint24`, and so on, in increments of 8 up to `uint256`.

If you use only `int`/`uint` in your code, these are aliases of `int256`/`uint256`. An `int`/`uint` data type is declared and assigned, as shown here:

```
int varInt = -25; //Signed integer, int is alias of int256
uint varUint = 25; //Unsigned integer, uint is alias of uint256
```

Regardless of the `intN` or `uintN` type you are using, its default value is 0. Here, N represents a number starting from 8 to 256, stepping 8. Meaning `intN` can be `int8`, `int16`, `int32`,..., `int256`. Similarly `uintN` means it can be `uint8`, `uint16`, `uint32`, ..., `uint256`.

Just like other mainstream languages, such as Java and C++, almost all operators in Solidity are supported with same operator syntax used in other languages. For developers, there is no special introduction required for each of these operators—their syntax is self-explanatory, as we assume that at least one programming language is known to the reader.

The following operators are supported by `intN`/`uintN`:

- **Comparisons or relational operators**: These are <=, <, ==, !=, >=, and >. All of these operators are a binary operator and their result evaluates to a `bool` type.
- **Bitwise operators**: &, |, ^, ~, <<, and >> are used to perform bitwise operations on integer types.
- **Arithmetic operators**: These are +, -, unary -, unary +, *, /, %, and ** (the exponentiation operator is specific to Solidity).

 A Solidity developer must know that the arithmetic operations performed on `intN`/`uintN` are prone to buffer overflow or underflow. It is not recommended to use arithmetic operators on the values of a data type directly. To avoid buffer overflow or underflow attacks, you should use the `SafeMath.sol` library provided by OpenZeppelin. These libraries ensure the safety of the arithmetic operations performed in your smart contract. You can refer `Chapter 9`, *Deep Dive into the OpenZeppelin Library* for more details.

Boolean value type

Solidity supports the `bool` data type that can store values such as `true` or `false`. A Boolean data type is declared and assigned a `bool` value, as shown here:

```
bool isInitialized = false;
```

If the value of a `bool` data type is not assigned, its default value will be `false`.

The following operators are supported on the `bool` data type—just like all of the other programming languages, unary and binary operators, including short circuit operators (`&&` and `||`), equal (`==`), not equal (`!=`), and unary negation (`!`) operators can be used on the `bool` data type:

```
require( ! isInitialized );  //Negation operator
bool isAllowed = isAdmin || isModerator; //Short circuit operator
if( getStatus() == isPending ) { //... }//Equal operator
```

The preceding code shows example usage of the operators allowed on the `bool` data type.

Address value type

The `address` data type represents the public address of an external account or a smart contract. An `address` data type stores 20 bytes value that represents an Ethereum address. An Ethereum address is always represented in hexadecimal and prefixed with `0x`. This makes the address of an Ethereum look something like `0xFf899Af34214B0D777bcD3c230EF637B763F0B01`.

An `address` data type is declared and assigned a value as shown in the following code:

```
address owner = 0xFf899Af34214B0D777bcD3c230EF637B763F0B01;  //Fixed value
without quotes
address sender = msg.sender; //Assignment
address emptyAddress = address(0);  //Sets the address to 0x0
address current = address(this); //Assigns current contract's address
require(msg.sender == owner);
```

All the `address` data type are assigned a default value of `0x0000000000000000000000000000000000000000`. This also known as `0x0`, `address(0)`, and `address(0x0)`.

There are many binary operators allowed on the `address` data type, including `==`, `!=`, `<`, `>`, `<=`, and `>=`. However, only equals (`==`) and not-equals (`!=`) operators are mostly used to compare two addresses.

An `address` type has the following member property and member functions. We will go through each of these properties and functions in detail:

- `balance`, `transfer()`, and `send()`: These are ether funds specific properties and functions
- `call()`, `delegatecall()`, and `staticcall()`: These are low level member functions

In the new Solidity version 0.5.0, a new enhanced address data type called address payable was introduced for security reasons. Using payable over address means that .send and .transfer functions will also be allowed to call on this address type. If a variable is only of the address type, it will not have the .send and .transfer functions available, but the other aforementioned member functions and member variables will be available. For example, msg.sender always returns an address of the address payable type, so that we can easily call msg.sender.transfer() to send ETH to the function caller.

Let's learn about each of the member properties and members functions available on the address type.

Reading a contract's ether balance

The address type has only one member property, that is, .balance. The syntax is as follows:

```
<address>.balance
```

.balance is a property; it is not a member function. This returns the current ether balance of an address as uint256 and is represented in wei.

The following is an example to get the current ether balance of the caller who initiated function call:

```
//This will return Ether balance of function caller.
uint256 balanceInWei = msg.sender.balance;
```

One thing to notice here is that if you are using the msg.sender.balance statement in your code, then it will return the current balance after deducting the maximum transaction fee required for the current call. For example, if an account X had a balance of 1 ether, and X initiates a transaction (setting the maximum transaction fee as 0.05 ether) containing msg.sender.balance in the function call to check the balance of the caller account, the following is how it will be calculated:

$$msg.sender.balance = BalanceBeforeTransaction - MaxTransactionFeeOnPendingTx$$

As per the preceding formula, balanceInWei will be initialized with 0.95 ether:

```
balanceInWei = 1 ether - 0.05 ether = 0.95 ether
```

Let's say that the preceding call successfully finishes, and account X gets 0.01 ether gas refunded, as that gas was not utilized. Hence, account X's actual current balance will be 0.96 ether.

Sending ether using transfer

The syntax for using `transfer` is as follows:

```
<address>.transfer(uint256 amount)
```

`.transfer` is used to transfer ether units in wei to another external account or contract account. In the call, `address` is the public address of external account or contract account, and the `amount` parameter is always specified in wei. The transfer method call fails if there is some exception, or failure occurred, perhaps because of running out of gas. It is recommended to use `transfer` over the `address.send` call.

Sending ether using send

The syntax for using `send` is as follows:

```
<address>.send(uint256 amount) returns (bool)
```

The `.send()` function is used to send ether units in wei to another external account or contract account. This is same as the `address.transfer` function call. The only difference is that this call returns a `bool` result to indicate that the `.send` operation succeeded or not. The execution does not stop when the `.send` call fails, and the return value is not validated.

If the `.send` operation fails for whatever reason, it will always return `false` as a result, but the call itself will not throw and revert the transaction on its own.

It is the developer's responsibility to check for the return value and take action accordingly. There are high chances of failure to check for the return value at the time of development of contract, hence the `.send` call is not recommended to be used.

To avoid these problems, it is better to use `.transfer` function calls instead of `.send`.

Understanding call and delegatecall functions

To call a function of another contract or to call a library function, you can use these low-level `call` and `delegatecall` function calls. These functions take an arbitrary number of arguments of any type that Solidity supports. Each of the arguments is padded to 32 bytes (padded with 0 zeros) and concatenated to create `payload` or transaction data.

The first argument you provide in these calls will be encoded to exactly 4 bytes and will not be padded. These 4 bytes define the signature of a function to be called on the target contract.

While writing the contract, you need to ensure that these two function calls should be avoided. If there is a need for your contract to use these functions, then extra care should be taken. Because these functions are low-level functions and they pass the control of the contract to the target contract, a malicious attacker can use the calls present in your contract to attack or drain funds from your contract.

Just to give you an example of how dangerous these calls could be—there was a very famous Parity multisig wallet contract hack that happened in 2017, in which an attacker used `delegatecall` to become the owner of the multisig wallet contract and withdrew ether worth more than $30 million.

Let's understand these two function calls individually.

The syntax for using `call` is as follows:

```
<address>.call(...) returns (bool)
```

This initiates a low-level `call` instruction with the given `payload` or transaction data. As explained earlier, it takes any number of arguments of any Solidity data type. Upon failure of the transaction, it returns `false` if not explicitly specified. This call forwards all remaining gas available to the target function call. As a developer, you can customize how much gas should be forwarded to the target call, to ensure that the target contract receives a sufficient amount of gas to successfully execute the transaction.

The syntax for using `delegatecall` is as follows:

```
address>.delegatecall(...) returns (bool)
```

This initiates a low-level `delegatecall` instruction. This behaves just like the `call` function works. The only difference is that it takes the code of the target/called contract and executes it in current caller contract. This means that it takes the Solidity code instructions from the target contract and executes it on the current contract context, which in turn modifies the storage variables, balance, and states present in current caller contract.

All of the calls to the library functions are initiated using `delegatecall` only because libraries define code; however, the storage of the caller contract is used.

The example code shows the `call` and `delegatecall` usage:

```
contract CallExample {
  //... {
  address otherContract = 0xC4FE5518f0168DA7BbafE375Cd84d30f64CDa491;
  string memory param1 = "param1-string";
  uint param2 = 10;

  //With multiple parameters
  require(otherContract.call("methodName", param1, param2));
  require(otherContract.delegatecall("methodName", param1, param2));

  //With signatures
  require(otherContract.call(bytes4(keccak256("methodName(string,
    uint256)")), param1, param2));
  require(otherContract.delegatecall(bytes4(keccak256("methodName(string,
    uint256)")), param1, param2));
  }
}
```

Understanding the staticcall function

The syntax for using `staticcall` is as follows:

```
<address>.staticcall(bytes memory) returns (bool, bytes memory)
```

This initiates a low-level `staticcall` instruction with a given `payload` or transaction data and returns a Boolean condition along with the return data. Upon failure of the transaction, it returns `false`. Just like the `call` instruction, if gas is not specified explicitly, it forwards all of the remaining gas to the target contract. In the contract itself, you can adjust the amount of gas units to forward to target contract.

As we have seen with the `call` and `delegatecall` low-level instructions, state modifications in the target contract or in the current contract are allowed. However, with the `staticcall` instruction, state modification will not be done on either the target called contract or the current contract.

The `staticcall` opcode was introduced in the Ethereum network upgrade release named the *Byzantium fork* on 16th October, 2017. The `staticcall` opcode is available only in the assembly language instructions of Solidity 0.4.25. The `staticcall` function is not explicitly available for use in Solidity version 0.4.25, but it is now available in Solidity version 0.5.0 and onward.

Adjust gas for a transaction using gas

While explaining `call`, `delegatecall`, and `staticcall`, we learned that the gas adjustments are allowed with these calls. This means that the developer can change the value of the gas units that they would want to forward to the target contract call. We can adjust the gas units using the `.gas()` function.

As a developer, you must know what kind of function call you are going to make to the target contract and how many instructions or pieces of code it will execute. After understanding the behavior of the target function call, you can decide and explicitly adjust the gas units. One thing to notice is that the gas units must be sufficient enough to successfully execute all of the instructions present in the target contract. For example, imagine that the gas requirement of your target contract execution is between 10,000-11,000 gas units and you have explicitly specified 10,500 gas units using the `.gas()` function. This could be dangerous because it is possible that some of your calls will always fail. To avoid this situation, add some extra gas units on top of your maximum gas requirement. Example usage of `.gas()` is shown in the following code, where it is forwarding 1 million gas units to `otherContract` to execute the `methodToCall` function on it:

```
otherContract.call.gas(1000000)("methodToCall", "param1");
```

Forwarding ether to another contract

Just like with `.gas()`, gas units can be forwarded to a called contract. Similarly, the ether value can be forwarded or sent to the called contract. This is done using the `.value()` method call. As shown in the following code, the current contract initiates the `methodToCall` function call on `otherContract` and forwards 1 `ether` to it. As `methodToCall` is going to receive 1 `ether`, it must be a `payable` function, as follows:

```
otherContract.call.value(1 ether)("methodToCall", "param1");
```

Notice that in the preceding code, to transfer the ether amount to `otherContract`, the current contract must have ether balance.

The `.gas()` option is available on both the `call` and `delegatecall` methods, while the `.value()` option is not supported for `delegatecall`, as follows:

```
contract GasExample {
    constructor () public {
        address otherContract = 0xC4FE5518f0168DA7BbafE375Cd84d30f64CDa491;
        /* gas adjustments */
        require(otherContract.call.gas(1000000)(
                "methodName", "param1"));
        require(otherContract.delegatecall.gas(1000000)(
```

```
        "methodName", "param1"));

    /* Wei forwarding using value() */
    require(otherContract.call.value(1 ether)(
            "methodName", "param1"));
    //.value() not supported on delegatecall, Compilation error
    //require(otherContract.delegatecall.value(1 ether)(
    //       "methodName", "param1"));

    /* Using gas() and value() together */
    require(otherContract.call.gas(1000000).value(1 ether)(
            "methodName", "param1"));
    //This is also valid
    require(otherContract.call.value(1 ether).gas(1000000)(
            "methodName", "param1"));
    }

}
```

Solidity version 0.4.25 also has the `callcode` method supported for the `address` data type, but it was deprecated. However, in Solidity version 0.5.0, `callcode` has been completely removed and is not supported anymore. Due to this reason, we have not covered `callcode` in this chapter.

Changes in Solidity version 0.5.0

The signatures of the `call`, `delegatecall`, and `staticcall` functions have changed in Solidity version 0.5.0 to improve security and provide return values. Prior to Solidity version 0.5.0, these function calls could take an arbitrary number of arguments of any Solidity data type and return only a single `bool` variable to notify the user about the success of the function call. However, they now accept only the `bytes memory` argument (also called a `payload`), and they return two variables—`bool success` and `bytes memory returnData`.

The `bool success` return value is to let the caller know whether the function call succeeded or not. It is recommended to use `require(success);` only after the low-level calls, to ensure that, if current transaction also fails, these low-level calls also fail while executing on another contract, as shown in the following example code. The `bytes memory returnData` return type returns the data returned by the target function call.

Look at the calls to these functions shown in the following code—first, you need to build `payload` for the transaction and then make the specific function call along with `payload`:

```solidity
pragma solidity ^0.5.0;

contract SpecialFunctions {
    function specialCalls() public {
        address otherContract = 0xC4FE5518f0168DA7BbafE375Cd84d30f64CDa491;
        bytes memory payload = abi.encodeWithSignature(
            "methodName(string)",
            "stringParam");

//Takes only bytes memory as argument
//Returns bool success, bytes returnData
        (bool successCall, bytes memory returnDataCall)
            = address(otherContract).call(payload);
        require(successCall);

        (bool successDCall, bytes memory returnDataDcall)
            = address(otherContract).delegatecall(payload);
        require(successDCall);

        (bool successSCall, bytes memory returnDataSCall)
            = address(otherContract).staticcall(payload);
        require(successSCall);
    }
}
```

Fixed size byte arrays

A **fixed size** array means the array that does not grow or shrink in terms of its capacity. Solidity supports the `byte` data type that represents a value of a byte (8 bits). There are some predefined byte arrays data types readily available in Solidity: `bytes1`, `bytes2`, and so on, up to `bytes32` (stepping 1). Each of these is a `byte[]` arrays type; for example, `bytes2` is a `byte[2]` array of size 2, `bytes3` is a `byte[3]` array of size 3, and so on for `bytes32` is a `byte[32]` array of size 32. `bytes1` is a special one, as it is an alias of `byte`.

As each of `bytes1`...`bytes32` is an array of `byte[]`, their values are accessible with the index. For example, if there is a variable, b, which is of the `bytesN` type, then b[k] returns the kth byte of the `bytesN` array, where k represents the index such that $0 <= k < N$, and N represents the bytes size, such that $1 <= N <= 32$.

All `bytesN` types are iterable using the loop because their byte elements are accessed using an index. Random access is also possible using an index.

Member property .length: These types of array have a .length member property, which returns the size of the byte[] array. You cannot modify .length as these are fixed length byte arrays.

All default bytesN type have a default value of 0, but if you get the data using the Remix IDE, it will be shown in hexadecimal representation as 0x<NumberOfZeros>, where *NumberOfZeros* = N * 2. For example, in the case of bytes4, its default value is 0, and in hexadecimal, it is 0x00000000.

The following are the operators supported on bytesN:

- **Comparisons operators**: These are <=, <, ==, !=, >=, and >. All are binary operators that evaluate to the bool type.
- **Bitwise operators**: &, |, ^, ~, <<, and >> are used for bitwise operations.

Dynamically sized byte arrays

In any programming language, you will need the support of strings for some specific needs. For string and dynamic length data, the Solidity language supports two types of dynamically sized byte arrays—the bytes and string data types. Neither are a value type, which means that their complete values are not copied when they are assigned to a new variable or passed to a function as an argument. However, their references are copied to a variable and arguments when passed to other functions.

Understanding the bytes data type

The bytes data type in Solidity is a dynamically sized byte[] array. As it's dynamic sized, the length of this type can grow and shrink. The bytes type is a bit different from bytes1, bytes2, ... bytes31, and bytes32 types (stepping 1), as bytes stores tightly packed data, whereas bytesN does not.

The bytes type variables are initialized with an empty string. If you check with the Remix IDE, it will return 0x, meaning a byte[] array with a length of 0.

There is no operator supported directly on bytes, but sometimes you might need to compare the two bytes variables. To check equality, you can compute the keccak256 hash of both the variables and perform a comparison, like that shown in following code example:

```
bytes a = hex'61';
bytes b = hex'61';
bool isEqual = keccak256(a) == keccak256(b);
```

`bytes` is another form of `byte[]` arrays, so the `length` and `push` member functions are available on the `bytes` type. Also, this is a dynamic type, so increasing and decreasing the size of the array is possible via `.push` and `.length--`. As it is an array type, random access is possible with the `bytes` type using an index.

Understanding the string type

The `string` data type is dynamically sized, just like `bytes`. However, the string type does have support for UTF-8 encoded characters. But on the other hand, the `string` type does not have `.length` and `.push` member functions available. `string` types are not iterable and do not have random access available.

The `string` data type variables are assigned by a default empty string, if not initialized.

The `bytes` type should be used for any arbitrary length of raw data, and the `string` type for arbitrary length UTF-8 string data. If your string length is known or you can limit the data in smaller size, then `bytes1` to `bytes32` are preferred, because they are cheaper in terms of gas consumption.

Passing functions using function types

Just as in C/C++, you can pass a function as an argument to another function, and the same can be done in Solidity. These are called `function` types. The variables of `function` types will be assigned when you pass a function as an argument to another function that receives a `function` type argument. In the same way, you can return a `function` type from a function.

The syntax to define `function` types is as follows:

```
function (<parameter types>) {internal|external}
[pure|constant|view|payable] [returns (<return types>)]
```

In the preceding syntax, the keywords within the square brackets, `[...]`, are optional. As you can see in the code, the `[returns (<return types>)]` syntax is also optional—it can be removed if there is no return type present in the `function` type.

The functions defined in the contract are by default `public`. However, the `function` types are by default `internal`. For visibility refer `Chapter 3`, *Control Structures and Contracts*.

There are two types of `function` type available—`internal` and `external`. Let's look at those individually. Before that, we'll have a look at the function selectors—a way to get `function` types.

You can use the `constant` keyword for functions until version 0.4.25. However, in version 0.5.0 onward, you cannot create `constant` functions. Instead of `constant`, you should use `view` for functions.

Get a function type with function selector

To get the `function` type and pass it as an argument to another function, you can use the function name itself. For internal functions, you can use the function name itself, such as `functionName`, and for external functions, use `this.functionName`. You can see an example of it in the *Using internal function types* and *Using external function types* sections.

All of the public and external functions have special members available, called `selector`, which returns the first 4 bytes of the function signature as `bytes4`. You can use `selector`, as shown in the following example code:

```
contract SelectorExample {
    //Returns first 4 bytes of method signature 0x2c383a9f
    function method() public pure returns (bytes4) {
        return this.method.selector;
    }
}
```

Using internal function types

As the name suggests, this kind of `function` type are for internal use within the contract. These internal functions can be either library internal functions or functions inherited from the super contract. All of these functions must be executable internally.

The following example that shows how to use internal `function` types:

```
library ArrayIteratorLib {
    function iterate(function(uint) pure returns (uint) _skipFn)
    internal pure returns (uint[] tempArr) {
        tempArr = new uint[](10);
        for(uint i = 0; i < 10 ; i++) {
            tempArr[i] = _skipFn(i + 1);
        }
    }
}

contract SkipContract {
    function skip1(uint _i) internal pure returns(uint) {
        return _i * 1;
    }
```

```
function skip2(uint _i) internal pure returns(uint) {
    return _i * 2;
}

function getSkipFunction(uint _fnNumber) internal pure returns
(function(uint) pure returns(uint) ) {
    if(_fnNumber == 1)
        return skip1;
    else if(_fnNumber == 2)
        return skip2;
}

//Returns Array[1,2,3,4,5,6,7,8,9,10]
function getFirst10WithSkip1() public pure returns (uint[]){
    return ArrayIteratorLib.iterate(getSkipFunction(1));
}

//Returns Array[2,4,6,8,10,12,14,16,18,20]
function getFirst10WithSkip2() public pure returns (uint[]){
    return ArrayIteratorLib.iterate(getSkipFunction(2));
}
}
```

Using external function types

External `function` types require an address on which the function can be called. Also required is the function signature, this represents which function is to be called on the target address.

The following code shows an example that uses external `function` types:

```
contract OraclizeService {
    address authorized = 0xefd8eD39D00D98bf43787ad0cef9afee2B5DB34F;
    modifier onlyAuthorized() {
        require(msg.sender == authorized);
        _;
    }
    QueryData[] queries;
    struct QueryData {
        bytes currency;
        function(uint, bytes memory)
            external
            returns (bool) callbackFunction;
    }
    event NewRequestEvent(uint requestID);

    function query(
```

```
        bytes _currency,
        function(uint, bytes memory) external returns(bool) _callbackFn
    ) public {
        //Registering callback
        queries.push(QueryData(_currency, _callbackFn));
        emit NewRequestEvent(queries.length - 1);
    }

    function reply(uint requestID, bytes response) public onlyAuthorized {
        require(queries[requestID].callbackFunction(requestID, response));
        delete queries[requestID]; //release storage
    }
}

contract OracleUser {
    modifier onlyOracle {
        require(msg.sender == address(oraclizeService),
            "Only oracle can call this.");
        _;
    }
    // known contract address of Oraclize Service
    OraclizeService constant oraclizeService =
        OraclizeService(0x611B947ec990Ba4e1655BF1A37586467144A2D65);
    event ResponseReceived(uint requestID, bytes response);

    function getUSDRate() public {
        oraclizeService.query("USD", this.queryResponse);
    }

    function queryResponse(uint _requestID, bytes _response)
    public onlyOracle
    returns (bool) {
        // Use the response data
        //...
        emit ResponseReceived(_requestID, _response);
        return true;
    }
}
```

In the preceding code address, 0xefd8eD39D00D98bf43787ad0cef9afee2B5DB34F is
used, which represents the Oraclize's contract address, which is authorized to call some
functions on the preceding contract.

Solidity reference types

Complex types such as `string` and `bytes`, which do not always fit into 256 bits in a variable, have to be handled more carefully than the other value types. Copying reference types is quite expensive and the developer can specify where these reference types should be stored. As shown in the following example, `string` is stored in `memory` temporarily:

```
string memory param1 = "Hello World !";
```

Similarly, the `storage` keyword can be used to store the variable data in Ethereum storage.

Understanding variables' data locations in Solidity

In Solidity, all of the state variables and local variables have a data location specified by default, or you can explicitly define a chosen data location. These data locations are called memory and storage. You can override the default data location by explicitly specifying along with the declaration of the variable. You must define a data location when using complex data types such as arrays, bytes, and structs. Following is the definition for memory and storage:

- **Memory**: Variables defined with a memory data location are short-lived and not stored on the blockchain. All of the function parameters and return types are by default set to the memory data location.
- **Storage**: Variables defined with storage data location are persisted in blockchain along with your contract. All of the state variables of a contract are by default set to storage data location. Also, the local variables you use in functions are set to storage.

The data location changes how the assignment of the variables will behave. If you assign a variable using memory data location to another variable that is also using the memory data location, then the data reference is shared. This means that if you make a change to the data in the first variable, it will also be reflected in another variable. In the same way, if storage is assigned to storage variables, they share the reference and any data change done by first will also be reflected in the second, which in turn changes the data in the persistent storage of the blockchain.

In all Solidity versions prior to 0.5.0, for complex data types, there is no explicit need to specify data locations, as they take the default data location. However, in version 0.5.0, data locations must be explicitly specified.

Using arrays in Solidity

In Solidity, arrays can be fixed size or dynamic size. We have already seen in the previous sections, *Fixed size byte arrays* and *Dynamically sized byte, arrays* of this chapter that `bytes` and `string` are special types of array.

You can set the visibility of an array to `public` and have Solidity create a getter function for it automatically. As you can see in the following `ArraysExample` contract, the `owners` array is set to `public`. Anyone would be able to access the owner's array outside of the contract as well, even without using any function.

Creating arrays using `new`—you can create variable length, in memory arrays using the `new` keyword. It is not possible to resize these memory arrays. You can create memory arrays as shown in this example:

```
address[] memory owners = new address[](10);
```

The following members functions are available on an array:

- `push`: This appends a new element at the last position in an array. You can use the `push` function call, along with a new value to be pushed. This function is applied to dynamically sized arrays only. This function also returns the new length of the array. It increases the size of the dynamic array. The following is the example contract that shows how to use the `push` function call:

```
contract ArraysExample {
    //Dynamic Array
    address[] public owners;
    constructor(address[] _owners) public {
        for(uint i = 0; i < _owners.length ; i++) {
            uint newLength = owners.push(_owners[i]);
        }
    }

    function removeLast() public {
        //Check to ensure that array has element
        //Without this check, .length will have integer underflow.
        require(owners.length > 0);
        //Removes the last element from dynamic array
        owners.length--;
    }
}
```

- length: Arrays have a `length` member, which returns the current size of the array. Dynamic arrays can be resized in storage by changing the `.length` member.
  Using `.length--`, you can remove the last element from an array. See the preceding example method, `removeLast()`, which removes the last element from the dynamic array. There is no other way possible in Solidity version 0.4.25 to remove an element from arrays. However, Solidity version 0.5.0 has support for the `.pop` function to remove the last element from a dynamic size array.

One thing to notice here that the `length` member can underflow when `.length--` is called and there is no element present in the array, so make sure that you have the `require(array.length > 0);` condition. Check this before calling `.length--`.

Creating a key value map using mapping

You can declare a mapping in Solidity using:

```
mapping(KEY_TYPE => VALUE_TYPE)
```

The `KEY_TYPE` cannot be a mapping, a dynamically sized array, a contract, `enum`, or `struct`. The remaining data types can be used for `KEY_TYPE`. The `VALUE_TYPE` can be any type; even a mapping is allowed.

The data used for `KEY_TYPE` is not stored in a mapping—its `keccak256` hash is generated and persisted, which in turn used to look up the value. Mappings do not have a length member to know their current size; hence, you cannot iterate a mapping.

For the mappings whose visibility is `public`, Solidity creates a getter function automatically. The `KEY_TYPE` will become a required parameter for the getter function, and it will return `VALUE_TYPE` data present in mapping.

The following example contract shows how to create, update, and get values from a mapping:

```
contract MappingExample {
    mapping(address => uint) public balances;

    function update(uint newBalance) public {
        //Adds a new mapping if not present.
        //Updates the new value if entry already present.
        balances[msg.sender] = newBalance;
```

```
        }

        //Increases balance by _increaseBy numbers
        function increaseBalance(uint _increaseBy) public {
            update(balanceOf(msg.sender) + _increaseBy);
        }

        function balanceOf(address _user) public view returns (uint) {
            //Gets the entry from mapping
            return balances[_user];
        }
    }
```

Resetting variables using the delete keyword

The delete keyword is used to reset the values of some data type and complex types such as arrays and structs. For example, if delete is used on int/uint, its value will be set to 0, which is the default value of the data type:

```
uint a = 1;
function reset() public {
    delete a; //resets the value of a = 0
}
```

In the same way, delete can also be used to reset the array types, where it assigns a dynamic array of length 0 if the array is a dynamic array. If an array is a static array, it resets all elements and keeps the length of the array same.

When the delete keyword is used for a struct, it resets all of the variables of the struct to their respective default values.

For mappings, you can reset a value by calling delete on its key. One thing to notice here is that delete a behaves like an assignment to the a variable; it restores the default value of the type.

Assigning variables with units

In Solidity, two types of units, ether and time units, are supported for literal numbers.

Ether units supported to simplify calculations related to ether amount. All of the ether unit amounts are converted into wei. Time units are supported to simplify calculations related to time. All of the time unit numbers are converted into the number of seconds in Unix epoch time.

Specifying ether amounts using ether units

A literal number can take a suffix of **wei**, **finney**, **szabo**, or **ether** and use it to convert between the sub-denominations of ether, where ether currency numbers without a suffix are assumed to be in **wei**:

Suffix example	In wei
1 wei	1
1 szabo	1e12
1 finney	1e15
1 ether	1e18

Supported units for time

Suffixes such as `seconds`, `minutes`, `hours`, `days`, `weeks`, and `years` used after number literals, and these can be used to convert between units of time, where seconds are the base unit of each and units are converted naively in the following way:

- `1 == 1 second`
- `1 minute == 60 seconds`
- `1 hour == 60 minutes`
- `1 day == 24 hours`
- `1 week == 7 days`
- `1 year == 365 days`

You should not rely on the aforementioned suffixes if you need some calendar-specific calculations, such as taking into account leap years. These units do not take care of leap years or leap seconds. Because of this reason, the `years` suffix has been deprecated.

Global special variables and functions

In Solidity, there are some global variables and functions such as `block`, `msg`, `tx`, and `gasleft`. These functions and variables are accessible globally in all functions and modifiers of your contract. Any information related to the current block, transaction, transaction sender, and transaction initiation parameters can be accessed using these variables and functions.

Using block and transaction properties

Special variables such as `block`, `msg`, `now`, and `tx` are globally available in all of your contracts and libraries. Your transaction execution sometimes needs information related to the sender/initiator of the transaction. In some specific calculations related to time, you would need the current timestamp to perform some operations.

Let's understand each of these variables in detail.

Getting block information using the block variable

The `block` variable gives all of the information related to a block. It provides the details of the current block in which your current transaction is being executed. It has some members available that you can use to get specific information:

- `block.coinbase`: This provides the miner's EOA address of the current block.

- `block.difficulty`: This provides the current difficulty level of the network as a `uint256` value, which is used by the miners to solve the **Proof of Work (PoW)** puzzle.

- `block.gaslimit`: This provides the total gas limit of the current block as a `uint256` value. At the time of writing this book, 8 million gas is allocated to each block on Ethereum mainnet.

- `block.number`: This provides the current block number, which is generated for the current block as a `uint256` value. Block numbers are generated consecutively.

- `block.timestamp`: This provides the timestamp when the block is generated, as a `uint256` value. The timestamp is in seconds since Unix epoch.

Getting sender transaction info using the msg variable

The `msg` variable provides the information related to the transaction sender and some transaction data:

- `msg.data`: This provides the full call data, represented as bytes, which is sent by the transaction initiator. The data includes the first 4 bytes of method-hash to be executed, followed by the arguments of the method.
- `msg.sender`: This provides the sender of the message's current call as the address. When an external contract call is initiated from a contract, `msg.sender` will return the contract address of the caller contract.
- `msg.sig`: This provides the first 4 bytes of the function signature, to be executed as `bytes4`.
- `msg.value`: This provides the amount of wei sent along with the transaction, as a `uint256` value.

Getting the current time using the now variable

The `now` variable provides the current block timestamp as a `uint256` value. This is an alias for `block.timestamp`.

You should not rely on the correctness of `block.timestamp` or `now`, because these timestamps can be manipulated by the miners to some extent. These should not be used to generate random numbers. Random numbers will be deterministic if you use timestamp values to try to generate randomness, and miners would be able to attack your contract.

Getting transaction info using the tx variable

The `tx` variable provides transaction-related information:

- `tx.gasprice`: This provides the gas price of the transaction set by transaction initiator, as a `uint256` value.
- `tx.origin`: This provides the original sender of the transaction from the full call chain as an address. As all of the transactions are always initiated from an EOA, we can say that `tx.origin` will always return an address of an EOA. Therefore, the use of `tx.origin` is not recommended. Refer `Chapter 14`, *Tips, Tricks, and Security Best Practices* for more details.

Special functions

For block and transaction-related information, only two globally available functions are available:

- `blockhash(uint blockNumber) returns (bytes32)`: This provides the hash of the given block number as `bytes32`. This function will only work for the most recent 256 blocks, excluding the current block. Hence, use this function carefully.
- `gasleft() returns (uint256)`: This provides the remaining gas left in the transaction as a `uint256` value.

Application Binary Interface encoding functions

Solidity provides many global functions to encode (including packed encoding) the different data types of Solidity. This encoded data is used to create `payload` data that can be sent to function calls for the external contract calls. These are also used to generate unique hashes of different values.

The following are the different functions available:

- `abi.encode(...) returns (bytes)`: ABI encodes the given arguments. Arguments can be of any type. It returns the encoded data as `bytes`.
- `abi.encodePacked(...) returns (bytes)`: This performs packed encoding of the given arguments. Arguments can be of any type. It returns the packed encoding of the data as `bytes`.
- `abi.encodeWithSelector(bytes4 selector, ...) returns (bytes)`: ABI encodes the given arguments. The first argument takes a function selector and from the second onward it takes any data type. It returns the encoded data as `bytes`.
- `abi.encodeWithSignature(string signature, ...) returns (bytes)`: This is equivalent to `abi.encodeWithSelector(bytes4(keccak256(signature), ...)`.

The following is an example `ABIEncodeExample` contract to show different encoding functions' usage and their behavior:

```
contract ABIEncodeExample {
    address addr; uint uInt; uint8 uInt8; uint16 uInt16;
    constructor() public {
        addr = 0x611B947ec990Ba4e1655BF1A37586467144A2D65;
```

```
        uInt = 20; uInt8 = 25; uInt16 = 30;
    }
    function testMethod(uint _a, uint8 _b) public view { //... }
    function encode() public view returns (bytes) {
        //Returns following concatenated
        // Prefix: 0x
        // addr  : 00000000000000000000
        //         000611b947ec990ba4e1655bf1a37586467144a2d65
        // uInt  : 00000000000000000000
        //         0000000000000000000000000000000000000000014
        // uInt8 : 00000000000000000000
        //         0000000000000000000000000000000000000000019
        // uInt16: 00000000000000000000
        //         000000000000000000000000000000000000000001e
        return abi.encode(addr, uInt, uInt8, uInt16);
    }

    function encodePacked() public view returns (bytes) {
        // Prefix: 0x
        // addr  : 611b947ec990ba4e1655bf1a37586467144a2d65
        // uInt  : 00000000000000000000
        //         0000000000000000000000000000000000000000014
        // uInt8 : 19
        // uInt16: 001e
        return abi.encodePacked(addr, uInt, uInt8, uInt16);
        //Packing of uint as per their size
    }
    // Each of the following calls will return the same data
    // Prefix  : 0x
    // selector: 13bd8af1
    // uInt     : 00000000000000000000
    //            0000000000000000000000000000000000000000014
    // uInt8    : 00000000000000000000
    //            0000000000000000000000000000000000000000019
    return abi.encodeWithSelector(this.testMethod.selector, uInt, uInt8);
  return
abi.encodeWithSelector(bytes4(keccak256("testMethod(uint256,uint8)")),
 uInt, uInt8);
  return abi.encodeWithSignature("testMethod(uint256,uint8)", uInt, uInt8);
 }
```

Error handling in Solidity

In Solidity code, when certain conditions are met, you can throw an exception and fail the transactions. You can check for method input argument's correctness and revert the transaction if arguments are not in order.

One important thing to note here is that when a transaction is reverted or failed, all of its state changes and variable modifications are reverted. Also, the gas consumption it has consumed till the revert of the transaction will be consumed and sent to miner, and the rest of the remaining gas will be refunded back to the transaction initiator. The error handling methods are as follows:

- `assert(bool condition)`: This invalidates the transaction if the condition is not met or `false`. It should be used for internal error verification. This is used to check for invariants during the development phase. Once rigorous testing is done on the code, you can remove all `assert()` statements. This should not be used in place of `revert/require`.

- `require(bool condition)`: This reverts the transaction if the Boolean condition is not met. It should be used to check for errors in input or external components. It is a recommended practice to check the method's arguments and preconditions with `require`, then move forward with other calculations and the logic of the code.

- `require(bool condition, string message)`: This is exactly the same as the `require` example outlined in the previous bullet point, but with a message. The only difference is that this provides an error message when a `require` condition fails. In your contracts, there could be many require checks. If a transaction fails and you want to find out what made it fail, it will be much harder to find out without an error message. A failed transaction with an error message makes it easier for developers to understand which condition wasn't met. However, this `require` condition with message consumes more gas, as it has to write the message in the logs.

- `revert()`: This aborts the execution of a transaction and reverts all state changes. It also refunds the remaining gas.

- `revert(string reason)`: This aborts the execution of a transaction and reverts all state changes. However, this is accompanied by an explanatory error message, which will be logged when this `revert` is executed..

You should make use of `require` as much as possible, instead of using `revert`. In development, use `assert` to check for invariants. Ensure to keep the error messages unique for each of the `require` and `revert` calls.

Cryptographic functions

Cryptographic functions can be used in contracts to calculate the `sha256` and `keccak256` hashes. The `ecrecover` function is used to verify that the hashes are signed by a specific address's private key:

- `keccak256(bytes memory) returns (bytes32)`: This is also called Ethereum SHA3. This computes the Ethereum-SHA-3 (Keccak-256) hash of the tightly packed arguments and returns it as `bytes32`.
- `sha256(bytes memory) returns (bytes32)`: This computes the SHA-256 hash of the input `bytes` and returns it as `bytes32`.
- `ecrecover(bytes32 hash, uint8 v, bytes32 r, bytes32 s) returns (address)`: Recovers the address associated with the public key from an elliptic curve signature or returns 0 on error.

When using `ecrecover` always check for return 0 and handle the condition accordingly.

Contract-related functions

Contract-related functions are available to get the address of the current deployed smart contract and to destroy a smart contract instance.

Get the contract address using this keyword

`this` is a keyword, used in the contract to get the current contract's type. `this` is also explicitly convertible to the `address` type. Here is some sample code explaining the usage of the `this` and `address(this)` keywords:

```
contract ThisExample {
  address public owner;
  ThisExample public instance;
  ThisExample public instanceConverted;
  address public currentContractAddress;
  address public currentContractAddressConverted;
  constructor() public {
    owner = msg.sender;
    instance = this;
    instanceConverted = ThisExample(this);
```

```
        currentContractAddress = this;
        currentContractAddressConverted = address(this);
    }
}
```

Destroying contracts using the selfdestruct function

`selfdestruct(address recipient)` is a globally available function. This is used to destroy an already deployed contract instance. Once the contract is destroyed on blockchain, all of its storage is released from blockchain.

The `address` argument of the function takes the recipient's address. If any ether balance is present in the contract, all of the ether balance will be sent to the recipient address.

As the `selfdestruct` function call releases storage from the blockchain, there are some gas refunds, which reduces the gas consumption and eventually leads to lower transaction fees.

An example of the `selfdestruct` function call is shown in the following code:

```
function kill() public {
    require(msg.sender == owner);
    selfdestruct(owner);
}
```

If the recipient address is a contract address with a payable fallback function defined, then the ether transfer triggered by the `selfdestruct` function does not initiate a call to the payable fallback function. Hence, the recipient contract will receive the ether balance of the destroyed contract silently, without calling the fallback function. You can refer Chapter 14, *Tips, Tricks, and Security Best Practices* for more details.

Destroying contracts using the suicide function

`suicide(address recipient)` is a globally available function. This is deprecated, so instead of using the `suicide` function, use the `selfdestruct` function.

The `suicide` function was available in Solidity version 0.4.25. However, in Solidity version 0.5.0, this function was removed. You should use the `selfdestruct` function only.

Topics for self-study

We have covered the basics of Solidity in this chapter. However, there are some topics that we haven't covered, which you can learn about with some self-study. For this, we recommend the following:

- Get an understanding of how string and hexadecimal literals are defined in Solidity. You can refer it online from Solidity documents at `https://solidity.readthedocs.io/en/v0.5.0/types.html#string-literals`.
- Understand how the conversions between different data types are supported in Solidity. You can refer it online from Solidity documents at `https://solidity.readthedocs.io/en/v0.5.0/types.html?#conversions-between-elementary-types`.

Summary

Solidity is a language that executes instructions by consuming gas, and this gas is paid to the Ethereum network's miners. For Solidity instructions to execute, consumers have to buy ether and pay for using the computation and storage resources of the Ethereum blockchain network. Before the innovation of the contract-oriented Solidity language, there was no other language that required crypto fuel to be paid for each instruction's execution. This makes Solidity a unique Turing-complete language.

We looked into the basics of the Solidity language in this chapter. We also covered the data types supported in Solidity. Solidity is a smart contract-oriented language that uses a very unique data type, called `address`. This `address` data type is the heart of the smart contracts. Almost all of the contracts that you will write will most likely use this data type. We also looked into the different global variables available in Solidity, which provide information about the sender, block, and transaction. Out of those global variables, `msg` is the one primarily used in contracts.

Solidity is a special language in the sense that it does not support null values for its data types. There is no special support for exception handling as of now, but maybe in the future, such support could be extended if Solidity programmers and developers need it and there is a use case for it, so watch this space.

In this chapter, we discussed the basics of Solidity language. As we move onto the next chapters, we will examine some of the more advanced topics of the Solidity language.

Questions

1. What is the difference between the `transfer` and `send` functions available on the `address` data type?
2. Why is `transfer` preferred over the `send` function call?
3. When should you use the `delegatecall` function?
4. How can you calculate the gas units to be set when doing transaction gas adjustments?
5. How can you generate a random number in Solidity?
6. What is the difference between the `abi.encode` and `abi.encodePacked` functions?
7. What is the difference between the `assert`, `revert`, and `require` functions?

Further reading

This book presents an overview on the topics that are a must for any developer learning the Solidity language. Nonetheless, this book also covers more advanced topics to help you to write production-ready Solidity-based smart contracts. If you need to focus only on the Solidity language, you can read *Solidity Programming Essentials* (`https://www.packtpub.com/in/application-development/solidity-programming-essentials`) by Ritesh Modi, from Packt.

3
Control Structures and Contracts

In this chapter, we will learn about the different control structures provided by Solidity that you can use while writing contracts, such as `for` and `while` loops, conditional statements, and much more. Then, we will move on to discussing different types of visibility modifiers for variables and functions, types of functions you can create, contract inheritance, abstract interfaces, and creating libraries.

The following topics will be covered in this chapter:

- Control structure keywords supported by Solidity
- Visibility modifiers applied to variables and functions of contracts
- Different types of functions and their behavior
- Solidity contract inheritance
- The use of abstract contracts
- Writing and using libraries in Solidity

Understanding Solidity control structures

A program doesn't always have a linear sequence of instructions. It may require some conditional and repeated code execution. This is why control structure keywords are supported in most languages.

Solidity supports many of the control structure keywords that are available in other languages, such as C and JavaScript. Conditions should be in parentheses (. . .) and code blocks should be in curly brackets { . . . }. The following are the control structures that Solidity supports.

The following are conditional control structures:

- `if...else` block: For conditional execution of the logic. You can also have nested `if...else` structures.
- `? :` operator: A ternary operator for conditional checking in a single statement.

The following are iteration control structures (loops):

- `while` loop: For creating `while` loops in Solidity
- `for` loop: Used to create `for` loops
- `do...while` loop: Also supports `do...while` loop constructs

The following are jump control structures:

- `break` keyword: To exit from the loop that is currently being executed on a certain condition
- `continue` keyword: To continue the loop for the next iteration upon a certain condition
- `return` keyword: To return one or more value from a function

When using any of the aforementioned control structures that need the condition to be defined, they need to be between parentheses, for example, (`<condition>`). This cannot be avoided. However, if you're only using a single statement code block, then you can avoid using curly brackets, as follows:

```
function initiateTransfer() public {
    if(isAllowed)
        transferFunds();
}
```

Let's learn how to return multiple values from a function.

Returning multiple values from function

In Solidity, from any type of function, you can return multiple values. You can do this by using the `return(value1, value2, ...)` statement or by directly assigning the value to the variable name that's used in the `returns()` statement.

The following is an example of returning multiple values from a function:

```
contract MultiReturn {

    function sum() public pure returns (uint) {
        //Receiving multi return from function
        (uint x, uint y) = getDataWithoutReturnStatement();
        return x + y;
    }

    //Example to return multiple values using return statement
    function getDataWithReturnStatement() internal pure
    returns(uint, uint) {
        return (1, 2);
    }

    //Example to return multiple values without return statement
    function getDataWithoutReturnStatement() internal pure returns(uint a,
    uint b) {
            a = 1;
            b = 2;
    }
}
```

As you can see, the `getDataWithReturnStatement()` function uses the `return` statement. However, the `getDataWithoutReturnStatement()` function is not using `return`; instead, it's assigning values to the variables specified in `returns()`.

Expression evaluation order

The expression evaluation order is not specified formally in Solidity. It doesn't specify the order in which the children of one node in the expression tree are evaluated. However, children are evaluated before the nodes are. It is guaranteed that short-circuiting for the Boolean expression is evaluated from left to right.

The following table shows the order of precedence of operators, with precedence 1 being the highest and precedence 16 being the lowest:

Precedence	Description	Operator
1	Postfix increment and decrement	`++, --`
	New expression	`new <typename>`
	Array subscripting	`<array>[<index>]`
	Member access	`<object>.<member>`
	Function-like call	`<func>(<args...>)`
	Parentheses	`(<statement>)`
2	Prefix increment and decrement	`++, --`
	Unary minus	`-`
	Unary delete operator	`delete`
	Logical NOT	`!`
	Bitwise NOT	`~`
3	Exponentiation	`**`
4	Multiplication, division, and modulo	`*, /, %`
5	Addition and subtraction	`+, -`
6	Bitwise shift operators	`<<, >>`
7	Bitwise AND	`&`
8	Bitwise XOR	`^`
9	Bitwise OR	`|`
10	Inequality operators	`<, >, <=, >=`
11	Equality operators	`==, !=`
12	Short circuit logical AND	`&&`
13	Short circuit logical OR	`||`
14	Ternary operator	`<conditional> ? <if-true> : <if-false>`
15	Assignment operators	`=, |=, ^=, &=, <<=, >>=, +=, -=, *=, /=, %=`
16	Comma operator	`,`

Since the specific order of evaluation is not specified, it is difficult to say how the calculations will be performed in a mathematical operation. For example, the preceding table shows that the same precedence, that is, 4, is applied to the multiplication, division, and modulo operations. This means that the evaluation of the following statement is unknown:

```
uint result = a * b / c;
```

Either the multiplication or the division will be done first. If the division is performed first over the small values, it is possible that there will be rounding errors. Therefore, always enclose the expression in brackets to ensure that the order of the evaluation as per your needs is maintained, especially for same precedence operators.

The preceding statement can be rewritten by including parentheses: (. . .):

```
uint result = (a * b) / c;
```

This ensures that the multiplication operation is done first, before division.

Solidity contracts

Just like the Java language, you can write your class definition and create as many objects as you want using the same class definition. Similarly, Solidity has smart contracts, in which you write the definition of the contract. When you deploy the contract on a blockchain, a public contract account address is generated and assigned to it. You can deploy as many contracts as you want with your contract definition on the blockchain, and each deployment would create new contract instances on the blockchain, each have unique contract addresses. Your application design determines whether you want to deploy multiple contracts of the same contract definition or not.

Note that contracts do not have the cron-like functionality necessary to auto-trigger a transaction from a contract at a given time. A transaction is always initiated from an EOA account.

There are two ways to create a new contract on a blockchain:

- Deploy the contract from outside of the blockchain using the contract definition and via the Remix IDE, truffle migration scripts, or the web3.js API.
- Create new contracts from an already deployed contracts. This is achieved by using the new keyword that's available in Solidity.

Let's look into each.

Cron is a software utility program that can schedule jobs or tasks to be executed in the future, at given times and intervals. You can find information about web3.js APIs and its usage on https://web3js.readthedocs.io.

Creating contracts

Once you have the contract definition ready and no compilation errors have been found, it can be deployed on the blockchain. You can use either the online Remix IDE, Truffle migration scripts, or the web3.js JavaScript API to deploy the contract on the blockchain. You can refer to Chapter 4, *Learning MetaMask and Remix* for more information.

When you deploy a contract from outside of the blockchain, it calls the constructor function, which is defined using the constructor keyword. The constructor is called only once, that is, at the time of deployment/creation. The constructor function is optional. You can also feed in some initial parameters to a contract via the constructor function by assigning arguments to it.

In the following code, we have defined two separate contracts:

```
//Contract with constructor
contract ConstructorExample {

    string public tokenName;
    string public symbol;
    address public owner;

    constructor(string _tokenName, string _symbol) public {
        owner = msg.sender;
        tokenName = _tokenName;
        symbol = _symbol;
    }
    //...
}


//Contract without constructor
contract NoConstructor {

    string public tokenName = "Sample Token";
    string public symbol = "SYMB";
    address public owner = msg.sender;
}
```

In the preceding example, the ConstructorExample contract is accepting arguments with its constructor. To deploy ConstructorExample, you must call the constructor and pass the arguments. On the other hand, the NoConstructor contract doesn't define a constructor, and so you can deploy it directly. By doing this, its state variables will be initialized with the values provided.

Creating child contracts using the new keyword

Solidity contracts also have the capability to create a new child contract of any definition. This is achieved by using the new keyword. Sometimes, depending on your application's design, you will want to create a child contract and get its address registered in the main contract.

One requirement for the new keyword is that, when it is being executed in your main contract, it should know the complete definition of the child contract that it is going to create.

The following is an example MainContract, which behaves as a register for every ChildContract it creates. When MainContract is deployed, it also deploys a ChildContract instance and assigns it to the ChildContract state variable. MainContract also provides two methods to create a ChildContract. These methods are createChildContract() and createChildAndPay():

```solidity
contract ChildContract {
    uint public id;
    uint public balance;
    constructor(uint _id) public payable {
        id = _id;
        balance = msg.value;
    }
}

contract MainContract {
    ChildContract[] public register;

    //ChildContract will be created when MainContract is deployed
    ChildContract public childContract = new ChildContract(100);

    constructor() public {
        register.push(childContract);
    }

    function createChildContract(uint _id) public returns(address) {
        ChildContract newChild = new ChildContract(_id);
        register.push(newChild);
        return newChild;
    }

    //Send ether along with the ChildContract creation
    function createChildAndPay(uint _id, uint _amount) public payable
    returns(address) {
        require(msg.value == _amount);
```

```
ChildContract newChild = (new ChildContract).value(_amount)(_id);
register.push(newChild);
return newChild;
    }
}
```

As you can see, along with `ChildContract` creation, you can transfer ether to `ChildContract` using the `.value()` function. However, gas adjustments are not possible for constructor creation. If the contract creation fails because of an out-of-gas exception or due to any other issue, it will throw an exception.

Using variable and function visibility

Solidity state variables and functions should have visibility defined, along with their declaration. This visibility defines how a function and a state variable will be accessible from within the contract, as well as from outside of the contract.

There are four different visibility modifiers, that is, `public`, `external`, `internal`, and `private`, present in Solidity:

- `public`: A function defined as `public` is accessible from both inside and outside of the contract. Solidity generates a getter function for `public` state variables, and these are also accessible from outside of the contract via these getter functions.

- `external`: A function defined as `external` is only accessible from outside of the contract. Internally, the function will not be accessible directly. To access an `external` function within the same contract, you would have to call `this.functionName()`. For state variables, you cannot use the `external` keyword.

- `internal`: A function or variable defined as `internal` is only accessible internally within the contract. Also, if Contract X inherits from Contract Y, which has `internal` functions or variables, Contract X can access all of Contract Y's `internal` functions and variables. When no visibility is specified, the variable would, by default, take `internal` visibility.

- `private`: A function or variable defined as `private` is only accessible internally within the same contract. A `private` variable or function is not accessible in derived contracts.

The following table explains different visibilities and how they apply to state variables and functions:

Visibility	Applies To	Accessibility
public	Both state variables and functions	Within the contract and outside of the contract
external	Only functions	Only from outside of the contract
internal	Both state variables and functions	Internally, also in derived contracts
private	Both state variables and functions	Internally, only within the same contract

To understand this better, let's look at the following example:

```
contract SuperContract {
    uint internal data;

    function multiply(uint _a) private pure returns (uint)
      { return _a * 2; }
    function setData(uint _a) internal { data = _a; }
    function externalFn() external returns (uint) { /*...*/ }
    function publicFn() public returns (uint) { /*...*/ }
}

contract VisibilityExample is SuperContract {
    function readData() public {
        //Following commented calls: error: not accessible
        //uint result = multiply(2);
        //externalFn();

        //Following calls: Allowed access
        data = data * 5; //variable accessible
        setData(10); //function accessible
        this.externalFn();
        publicFn();
    }
}

//Contract accessing VisibilityExample contract
contract ExternalContract {
    VisibilityExample ve = VisibilityExample(0x1);

    function accessOtherContract() public {
        //Following commented calls: error: not accessible
        //ve.setData(10);
        //ve.multiply(10);

        //Following calls: Allowed access
        ve.externalFn();
```

```
        ve.publicFn();
        ve.readData();
    }
}
```

As you can see, ExampleContract can call the externalFn(), publicFn(), and readData() functions.

If you have a public function that accepts an array as an argument, it is better to make that function external, as it is more efficient in terms of gas consumption.

Let's understand the visibility and accessibility by looking at preceding example. The following examples are shown in example code that follows in this chapter:

- The VisibilityExample contract inherits from SuperContract. Refer to the *Contract inheritance* section of this chapter for more information.
- The state variable data present in SuperContract is also accessible in VisibilityExample and can be modified directly.
- The multiply() function is not accessible in the VisibilityExample contract, as its visibility is private. This function is only accessible inside SuperContract.
- The externalFn() function is not accessible directly inside the VisibilityExample contract. However, you can access it by using this.externalFn().
- ExternalContract is deployed separately from the VisibilityExample contract. This means that both of the contracts have different contract addresses. However, ExternalContract can call the functions of the VisibilityExample contract.

A private function or state variable within a contract doesn't mean that no one will be able to see that function. Since the blockchain is public, so is the definition of the contract, and anyone will be able to see the function definition. However, private visibility is only used to restrict other contracts from calling or accessing it.

Next, let's have a look at getter functions for state variables.

Getter functions for state variables

If you have `public` state variables in your contract, the compiler will create getter functions for these automatically. Therefore, if you have already defined `public` state variables, you don't have to write getter functions explicitly for those variables. It isn't recommended to write getter functions for `public` state variables.

However, if you have the array as a `public` state variable, then you can access a single element of that array by providing an `index` as a parameter to the function.

Let's look at the following `GetterExample` contract, where we have two public state variables, `data` and `array`:

```
contract GetterExample {
    uint public data = 25;
    uint[2] public array = [10,20];
    //Overrides getter function of `data` state variable, if defined
    function data() public pure returns (uint) {
        return 15;
    }
    //Overrides getter function of `array` state variable, if defined
    function array(uint _i) public pure returns (uint) {
        return 60 + _i;
    }
}

contract ExternalContract {
    GetterExample ge = new GetterExample();
    function getData() public view returns (uint) {
        return ge.data();
    }

    function getArray(uint _index) public view returns (uint) {
        return ge.array(_index);
    }
}
```

As you can see, the `data` state variable will have its getter function created as `data()`, but for the `array` state variable, the getter function will be `array(index)`. You are allowed to override the getter functions.

Creating custom function modifiers

In Solidity, you can write function modifiers to check for any preconditions before executing a function. These modifiers are always inheritable in the inheritance hierarchy. You can also override the definition of the modifiers.

Every modifier must have _; present in its definition, which means that the body of the function where the modifier is used will be placed at this location. It tells the program to go back to where it was originally called to continue execution. The following is an example of the onlyOwner and logAtEnd modifiers being defined:

```
contract Ownable {
    address public owner;
    modifier onlyOwner() {
        require(msg.sender == owner);
        _;
    }
    constructor() public {
        owner = msg.sender;
    }
}

contract ModifierExample is Ownable {
    enum Status {INIT, PENDING, INPROCESS, PROCESSED}
    Status public status;

    event StatusChanged(Status status);
    modifier logAtEnd(Status _status) {
        _;
        emit StatusChanged(_status);
    }

    function changeState(Status _status) public onlyOwner
     logAtEnd(_status) {
        status = _status;
    }
}
```

Modifiers are mainly used to verify the argument's allowed values and for access controls. In the preceding code, we have an `Ownable` contract, which has the `onlyOwner` modifier; this checks whether `owner` is always the deployer of the contract. The `ModifierExample` contract inherited the `onlyOwner` modifier from the `Ownable` contract. The `changeState` function of the contract ensures that the method call is only allowed from the `owner`. When `owner` calls the function, then first state is changed and at the end log is emitted via `logAtEnd` modifier.

Creating constant state variables

Solidity supports the `constant` keyword for state variables and functions. The state variables that are defined as constant will have their values assigned at compile time and will not allow for any modifications at runtime. You cannot use blockchain data or block information such as `now`, `msg.value`, `block.number`, or `msg.sender` to assign constant state variables. However, you can use other built-in functions, such as `keccak256`, `ecrecover`, and `sha256`.

The following is the example contract defining constants:

```
contract ConstantExample {
    string public constant symbol = "TKN";
    uint public constant totalSupply = 10 ** 9;
    bytes32 public constant hash = keccak256(symbol);
}
```

You could also use the `constant` keyword for functions. However, this has been deprecated, and the `constant` keyword's use for functions was removed in Solidity version 0.5.0. Instead of using `constant`, you can use the `view` keyword. Refer the next section, where we discuss Solidity functions.

Understanding types of Solidity functions

Solidity functions take arguments and can return multiple return values. The function definitions start with the `function` keyword. Every function should have its visibility specified. If this is not defined in version 0.4.25, it will default to `public` visibility. In Solidity version 0.5.0, you must have visibility defined for a function; otherwise, you will get compilation errors.

The syntax that's used to define a function definition is as follows:

```
function functionName(<parameter types>) {internal|external|public|private}
[pure|view|payable] [returns (<return types>)]
```

The following is the syntax diagram for defining a function:

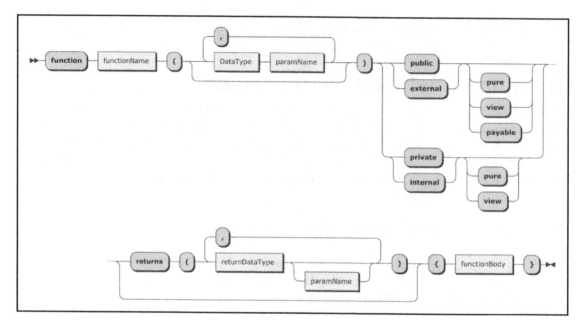

Syntax diagram for function definition

As shown in the preceding syntax diagram, a function can return multiple values using the `returns(value0, value1, ... , valueN)` syntax.

Using view functions to read state variables

A function can be a `view` function when it is not modifying any state variables of the contract. The compiler also throws a warning message during compilation in order to convert a function into a `view` function if it is not modifying any state variables. You should resolve those warnings and use the `view` modifier for the functions indicated by the compiler warning; this would reduce gas consumption and data would be returned instantly. The `view` functions are mainly used to return the current state of the contract.

The following code is getting the current state of the exchange:

```
contract ViewFuncExample {
    enum ExchangeStatus { Inactive, Active }
    ExchangeStatus status = ExchangeStatus.Active;

    function getCurrentState() public view returns (ExchangeStatus) {
        return status;
    }
    //Rest of the code
}
```

In the preceding code, the getCurrentState() function is declared as a view function, which returns the current state of the status (enum ExchangeStatus) variable.

The view functions are a good way to reduce gas consumption because they are also accessible and return values for off-chain calculations. Off-chain accessibility means that it reads the latest state data of the contract from the blockchain via the view functions. These function calls do not consume gas as they are performed using message calls. The data is returned from blockchain instantly as the message calls do not require execution by miners and are not added in the block; the node you are connected to processes the message call and returns the value.

However, if you call a function that changes the contract state variables and this function internally calls a view function, it would also consume the gas of the transaction. This is why view functions are accessible both off-chain and on-chain. When called off-chain, they don't consume gas, but when they're called on-chain, they do. To understand this, let's look at the following example code:

```
function withdrawDividend() public {
    //Code here to update dividend paid status, so that user
    //cannot claim again
    //Send current dividend amount
    msg.sender.transfer(calculateDividend());
}

function calculateDividend() public view returns (uint) {
    return balanceOf[msg.sender] * dividendUnitPerToken;
}
```

As shown in the preceding example, we have a view function called `calculateDividend()`, which is allowed to be called off-chain. When this function is called off-chain, it does not consume gas and returns the dividend amount that the function caller is eligible to receive from the contract. However, when the user calls the `withdrawDividend()` function to withdraw their share of dividends from the contract, this call requires a transaction (on-chain). This transaction first updates the dividend paid status for the user, then internally calls to `view` the `calculateDividend()` function; now, this call to `view` the `calculateDividend()` function would also consume gas for its execution. Furthermore, the transaction transfers the correct dividend amount to the user.

The state variables that are defined as `public` have getter functions created for them automatically by the compiler. These getter functions are created as `view` functions.

You cannot mark a function as a `view` function if you are doing any of the following inside that function, because all these change state in the contract:

- Writing to a contract state variable
- Calling any function within the current contract or on another contract that isn't marked as `pure` or `view`
- Emitting any events
- Creating new contracts
- Using the `selfdestruct` call
- Sending ether to another contract
- Using any low-level calls that change the state variables
- Using an inline assembly that contains certain opcodes that make changes in the blockchain

Using pure functions to perform calculations

The `view` function only reads the contract state variables, but it doesn't make any modifications to them. On the other hand, the `pure` function doesn't read the state variables of the contract, also not make any modifications to the contract state variables.

These `pure` functions are also accessible off-chain; you can read and perform calculations using these functions off-chain. These off-chain calculations don't require gas to be paid. However, if a non-view or non-pure function call is made using a transaction, a `pure` function would also consume gas from the transaction to execute its instructions.

The following are the ways in which we can make state variables read calls. These are not allowed in a pure function:

- Reading any state variable defined in the contract.
- Using address(this).balance to read the ether balance of a contract.
- Reading any of the members of globally available variables such as msg, tx, and block. However, msg.data and msg.sig can be used in pure functions.
- Calling any function that isn't marked as a pure function.
- Using inline assembly opcodes that read the state variables.

You can perform the following in a pure function:

- An arithmetic calculation without reading any contract state variables.
- Use msg.data and msg.sig.
- Use the globally available function that does not make state changes, for example, the keccak256() and ripemd160() functions.

The following example contract shows a pure function:

```
contract PureFuncExample {
    function add(uint _a, uint _b) public pure returns(uint) {
        uint c = _a + _b;
        require(c >= _a);
        return c;
    }
}
```

In the preceding code, we are adding two unsigned integers, and also checking that the addition operation does not overflow.

Using the default fallback function

A Solidity contract may have only one fallback function defined. This is a special function that doesn't have a function name, doesn't take any arguments to the function, and also doesn't return any values from the function. The fallback function is executed when the data in the transaction is empty, as well as when there is no matching function identifier present in the contract.

The fallback function is defined as follows in the contract:

```
function() external {
    //function body
}
```

The fallback function could have any visibility specified, that is, up until Solidity version 0.4.25. However, version 0.5.0 onwards, only `external` visibility is allowed.

If the fallback function is `payable`, the contract can receive ether. It can also receive ether if it is sent to the contract address with or without data. If the fallback function doesn't exist in the contract, or if it isn't a `payable` function, ether that's sent to the contract address will be rejected and the transaction will fail.

Here, a `payable` fallback function that can accept any amount of ether that's sent to its contract has been defined. The function also tracks the amount it has received in the `weiReceived` state variable:

```
contract FallbackFunctionExample {
    uint weiReceived;

    function() external payable {
        weiReceived += msg.value;
    }
}
```

However, there are some exceptions. If ether is sent to any contract using the `selfdestruct` function call, the contract's fallback function will not be executed, even if the fallback function is defined as `payable`. Furthermore, this transaction will not be rejected/failed. This also applies when the contract address is receiving ether from mining block rewards as a coin base transaction.

Although the fallback function cannot receive arguments, it can still read `msg.data`, and extract and process the payload that's supplied with the transaction.

Overloading functions

In Solidity, overloading a function means that you can create a function that has the same name but different arguments and/or a different type. For function overloading, you have to ensure that the function signature is different for each overloaded function. Overloading also applies to inherited functions.

The function signature consists of the function name and its argument type in the order they are defined. Let's look at the following example:

```
contract FuncOverload {
    address owner;
    constructor() public {
        owner = msg.sender;
    }

    function transfer(address _to, uint _amount) public returns (bool) {
        return doTransfer(_to, _amount);
    }

    function transfer(address _to) public returns (bool) {
        return doTransfer(_to, 1 ether);
    }

    function transfer() public returns (bool) {
        return doTransfer(owner, address(this).balance);
    }

    function doTransfer(address _to, uint _amount) internal
    returns (bool) {
        _to.transfer(_amount);
        return true;
    }
}
```

In the preceding code, we have overloaded the transfer() function in three different ways.

Overriding function definition

In function overloading, you define different operations per function definition. In function overriding, you can override the function definition of an existing function, which means you can replace the function definition of an inherited function. If you need to execute the old definition of the function, you can use super.functionName.

Function overriding can only be done by using contract inheritance. You cannot perform function overriding in the same contract. As shown in the following example, the transfer function is already present in BasicToken.sol. However, in the FuncOverride contract, we have added more functionality to the transfer method.

Now, this method will be called when the contract is not paused:

```
import "openzeppelin-solidity/contracts/token/ERC20/ERC20.sol";
import "openzeppelin-solidity/contracts/lifecycle/Pausable.sol";

contract FuncOverride is ERC20, Pausable {

    //Overriding transfer function of BasicToken.sol
    function transfer(address to, uint256 value) public whenNotPaused
    returns (bool) {
        return super.transfer(to, value);
    }
}
```

The preceding contract uses OpenZeppelin library contracts. We will look into the contracts provided by the OpenZeppelin library in Chapter 9, *Deep Dive into the OpenZeppelin Library*.

Using emit and events for event logging

Events in Solidity are used to log information. Off-chain client applications can subscribe and listen for any events that are being emitted in the blockchain. These client applications use the **Remote Procedure Call (RPC)** interface to interact with the Ethereum nodes and are notified upon event emission.

By using the Solidity language, your contracts can emit events, but they cannot listen to or read the data of a triggered event. Events are always inherited from contracts, which means that you can emit events that are defined in your contract inheritance tree.

When you emit an event, its arguments are stored in the transaction log. These logs are related to the contract address and are stored in the blockchain's block. These events will be available forever since they are present in the block. However, this might change in future versions of Ethereum blockchain upgrades via a hard fork, which changes the fundamental rule and policy of the blockchain.

You can define up to three indexed arguments in an event. The indexed event argument forms the topic that is used to filter events quickly in off-chain client-side applications. A topic can only hold one word (32 bytes) in storage. The topic, which helps with finding events with given indexed parameters, is generated as follows:

```
keccak(EVENT_NAME+"("+EVENT_ARGS.map(canonical_type_of).join(",")+")")
```

In the preceding format, EVENT_NAME represents the name of the event, and EVENT_ARGS.map(canonical_type_of).join(",") joins all the indexed arguments that are present in the event. Here, canonical_type_of represents the canonical type of a given argument; for example, in uint indexed foo, uint256 is the canonical type.

Non-indexed arguments will be stored in the data part of the log. Here, the contract will emit a Deposited event when it receives ether:

```
contract EventExample {
    uint public balance;
    event Deposited(address indexed from, uint amount);

    function () external payable {
        balance += msg.value;
        emit Deposited(msg.sender, msg.value);
    }
}
```

In the Deposited event, the from argument is indexed, which will allow off-chain client applications to filter for the events with the given address and get notified when such an event is triggered on the blockchain.

Inheriting contracts

Just like object-oriented programming languages, Solidity contracts also support multiple inheritance. You can derive another contract by using the is keyword, as shown in the FuncOverride contract that we discussed in the *Overloading functions* section of this chapter.

The contract can access the following from its inherited contract:

- All of its modifiers. You can also override them.
- All public and internal functions. You can also override them.
- All public and internal state variables. Use these state variables directly.
- All events. You can emit these events.

When a contract inherits from another contract and a function is called, its most derived function definition will be executed from the inheritance hierarchy. However, you can also use the contract's name or the `super` keyword in order to use a specific function definition:

```solidity
contract Ownable {
    address public owner;
    modifier onlyOwner() {
        require(msg.sender == owner);
        _;
    }

    constructor() public {
        owner = msg.sender;
    }
}

contract ValueStorage is Ownable {
    uint public value = 2;
    function update() public onlyOwner {
        value += 1;
    }
    //...
}

contract ValueStorage1 is ValueStorage {
    function update() public {
        value *= 2;
        ValueStorage.update();
    }
}

contract ValueStorage2 is ValueStorage {
    function update() public {
        value *= 3;
        ValueStorage.update();
    }
}

contract InheritanceExample1 is ValueStorage1, ValueStorage2 {
}
```

When you deploy this contract, the InheritanceExample1 contract calls the update()
function on it. The value state variable will be set to 7. Because
the ValueStorage2 contract is mostly derived, the update() function will multiply 2 * 3 =
6 and then call the ValueStorage.update() function, which will calculate 6 + 1 = 7. Due
to this, the ValueStorage1.update() function is skipped.

Let's look at the same example, except this time, we will use the super keyword:

```
contract Ownable {
    address public owner;
    modifier onlyOwner() {
        require(msg.sender == owner);
        _;
    }

    constructor() public {
        owner = msg.sender;
    }
}

contract ValueStorage is Ownable {
    uint public value = 2;
    function update() public onlyOwner {
        value += 1;
    }
    //...
}

contract ValueStorage1 is ValueStorage {
    function update() public {
        value *= 2;
        super.update();
    }
}

contract ValueStorage2 is ValueStorage {
    function update() public {
        value *= 3;
        super.update();
    }
}

contract InheritanceExample2 is ValueStorage1, ValueStorage2 {
}
```

When you deploy `InheritanceExample2` and call the `update()` function, the `value` state variable will be set to `13`. The inheritance sequence, starting from the most derived contract, is as follows: `InheritanceExample2`, `ValueStorage2`, and `ValueStorage1`. Therefore, the following functions will be called in order:

- `ValueStorage2.update()`: Perform 2 * 3 = 6 and set `value`
- `ValueStorage1.update()`: Perform 6 * 2 = 12 and set `value`
- `ValueStorage.update()`: Perform 12 + 1 = 13 and set `value`

Passing arguments for the base constructor

In the inheritance hierarchy of the contracts, their constructors will be called as per the linearization rule. If any of the contracts need arguments to be passed it, it must supply those arguments from the derived contracts.

There are two ways you can pass arguments to the constructors of the base contract:

- Pass constructor arguments from the inheritance list.
- Pass constructor arguments from the derived contract's constructor.

You can use either approach—they are both easy to implement. Let's take a look at them:

```solidity
import "openzeppelin-solidity/contracts/token/ERC20/ERC20.sol";

contract DetailedERC20 is ERC20 {
    string public name;
    string public symbol;
    uint8 public decimals;

    constructor(string _name, string _symbol, uint8 _decimals ) public {
        name = _name;
        symbol = _symbol;
        decimals = _decimals;
    }
}

contract MyToken is DetailedERC20("MyToken", "MTKN", 18) {
}

contract MyToken2 is DetailedERC20 {
    constructor (
        string _name,
        string _symbol,
        uint8 _decimals
```

```
    ) DetailedERC20(_name, _symbol, _decimals) public {

    }
}
```

In the preceding example, the `MyToken` contract doesn't require its own constructor to be declared as it has directly passed arguments to the `DetailedERC20` contract from the inheritance declaration itself. Another way of doing this is shown in the `MyToken2` contract, where it inherits `DetailedERC20` and passes the arguments to the `DetailedERC20` contract.

By looking at the preceding example, we can see that the first approach should be used when we know the arguments before we deploy the contract. This is why you hardcode the arguments in the constructor definition itself. On the other hand, the second approach allows you to pass the arguments to the constructor of the `MyToken2` contract when arguments are not known and decided at the time of deployment.

Understanding inheritance linearization

Multiple inheritances can have many problems, such as the well-known **diamond problem**. Solidity uses the C3 linearization approach to force a specific order for inheritance. In Solidity, the order in which the contracts are specified as using the `is` keyword is important. The most derived contracts are the rightmost ones. For more information on C3 linearization and how it works along with an example, refer to Chapter 6, *Taking Advantage of Code Quality Tools*, the *Seeing inheritance dependencies* section.

When a function is defined in the contracts that are present in the inheritance hierarchy, the function is searched from the rightmost to the leftmost inherited contract in a depth-first search manner. The search stops when it finds its first match. It skips contracts that have already been searched.

Creating abstract contracts

In an abstract contract, there are only a few functions that don't have the function body defined. You cannot deploy abstract contracts alone. However, you can inherit the contract and provide the definition to each function that's declared in the abstract contract.

If a contract inherits from an abstract contract and doesn't provide the implementation of the function, the inheriting contract would also be considered an abstract contract. Hence, it is the developer's responsibility to ensure that all the functions of the abstract contract are defined. Also, developer tools such as Remix and Truffle would not allow your inherited contract to be deployed.

As shown in the following example, AbstractDeposit is an abstract contract since it has a depositEther() function, and no function body has been provided for it. However, the DepositHolderImpl contract is implementing the AbstractDeposit contract and provides the definition of the depositEther() method:

```
contract AbstractDeposit {
    function depositEther() public payable returns (bool);
}

contract DepositHolderImpl is AbstractDeposit {

    mapping(address => uint) deposits;

    function depositEther() public payable returns (bool) {
        deposits[msg.sender] += msg.value;
    }
}
```

There is another way you can make an abstract contract, and that is by setting the visibility of the constructor to internal. However, you can define methods for this abstract contract, as shown in the following code:

```
contract InternalConstructor {
    uint public value = 10;

    constructor () internal { }

    function setValue(uint _value) public {
        value = _value;
    }
    //No abstract function present
}
```

As in the preceding example of an InternalConstructor contract, the constructor is defined as internal. Hence, you would not be able to deploy this contract. You can use this technique when you want to define a contract with functions and features that would be used when inherited in other contracts. However, you can see that there is no abstract function present in this contract, hence you made the constructor internal.

Creating interfaces

You can define interfaces in Solidity using the `interface` keyword. These interfaces are very similar to abstract contracts, and they must not have any function definitions. They also have the following restrictions:

- All the functions that are defined in the interface must have external visibility.
- The constructor is not allowed.
- An interface cannot have any state variables defined.
- The interfaces cannot inherit from any other contracts or interfaces.

There are some differences between the Solidity 0.4.25 and 0.5.0 version interfaces. In version 0.4.25, you cannot define `enum` and `struct`. However, with version 0.5.0 onward, you can define them.

The following example shows an interface that's been defined in Solidity version 0.4.25, having a function declared without the body:

```
pragma solidity ^0.4.25;

interface ExampleInterface {
    function transfer(address _to, uint _amount) external returns (bool);
}
```

The following example shows an interface that's been defined in version 0.5.0:

```
pragma solidity 0.5.0;

interface ExampleInterface {
    enum Status {Pending, Inprocess, Processed}
    struct Data {
        address requester;
        uint amount;
        Status status;
    }
    function transfer(address _to, uint _amount) external;
}
```

In the preceding example, we can see that `enum Status` and `struct Data` are also defined. All the inherited contracts would be able to use this `enum` and `struct`.

Creating custom reusable libraries

The libraries in Solidity are just like contracts, but they are deployed only once and their code is reused in the calling contracts. You can define libraries using the `library` keyword. Calls to the library functions use the `DELEGATECALL` opcode, which means that when a function on a library is called by the contract, only the code of the library function is executed in the context of the calling contract, and the storage of the calling contract is used and modified by the library. The library can have `pure` and `view` functions, which will be accessible directly from the calling contract because they do not initiate `DELEGATECALL`. You cannot destroy a deployed library.

When a library is linked to a contract, you can see that library as the implicit base contract of the contract and can access the functions defined in the library just by using the name of the library and its functions. For example, with `Lib` being the name of the library and `fname` being the function name, you can call `Lib.fname()` directly from the calling contract. All the internal functions of the library are also accessible to the contract, just like how they're available with the inheritance hierarchy.

Once the library has been created and tested, you can reuse that library code in any of your contracts as many times as you want and can use the code for any of your other projects that contain Solidity contracts.

In the following example, we have created a `ControlledAddressList` library, which maintains a list of addresses and keeps a flag of each address to show their status as either enabled or disabled. By calling the `enable/disable` functions of the library contract, you can change the status of an address. Then, by calling the `isEnabled/isDisabled` functions, a contract can read the status of an address.

Library data is stored in a `Data` struct. As you can see, in the functions of the library, we are passing a `storage` pointer to the `Data`. This `Data` will be stored in the calling contract storage and will pass the reference of that `Data` to the library in order to perform the action on its storage:

```solidity
library ControlledAddressList {

    struct Data {
        mapping(address => bool) addresses;
    }

    function enable(Data storage self, address _address) public
    returns (bool) {
        require(_address != address(0));
        require(isDisabled(self, _address));
```

```
        self.addresses[_address] = true;
        return true;
    }

    function disable(Data storage self, address _address) public
    returns (bool) {
        require(_address != address(0));
        require(isEnabled(self, _address));
        self.addresses[_address] = false;
        return true;
    }

    function isEnabled(Data storage self, address _address) public view
    returns (bool) {
        return self.addresses[_address] == true;
    }

    function isDisabled(Data storage self, address _address) public view
    returns (bool) {
        return self.addresses[_address] == false;
    }
}
```

The preceding library can be used in many places where you would need a list of addresses:

- For whitelist purposes
- To maintain controlled access for a function or contract calls
- For **KYC** (short for **Know Your Customer**) status

As the calls to the libraries are DELEGATECALL, using msg.sender or msg.value will not change their value if they're used in the library; instead, they will still represent values of the calling contract or account.

Let's learn how to use the libraries in the contract.

Using libraries with – using...for directive

We saw how to create a library in the previous section. Now, let's see how we can use those libraries in the contracts and make a call to the functions of the library.

To attach a library to the contract, we use a special directive called `using X for Y;`, where `X` is the library and `Y` is the data type. By defining this directive, we are saying that `X` is a library, we take all the functions from the `X` library, and allow those functions to be called on the `Y` data type. For example, the `X` library has a function called `funcName()`; in that case, you would be able to call `Y.funcName()` on the `Y` datatype. When `funcName()` is called, the first parameter to the `funcName()` function is passed as the `Y` datatype.

Here is an example contract that's called the `TokenList` contract, which is using `ControlledAddressList`, which we discussed in the previous section:

```
import "./ControlledAddressList.sol";
import "openzeppelin-solidity/contracts/ownership/Ownable.sol";

contract TokenList is Ownable {

    using ControlledAddressList for ControlledAddressList.Data;
    ControlledAddressList.Data data;

    function add(address _token) public onlyOwner returns (bool) {
        return data.enable(_token);
    }

    function remove(address _token) public onlyOwner returns (bool) {
        return data.disable(_token);
    }

    function isPresent(address _token) public view returns (bool) {
        return data.isEnabled(_token);
    }
}
```

The `TokenList` contract is maintaining the list of ERC20 token addresses. These ERC20 addresses can only be added or removed by the owner of the contract. As we can see, `using ControlledAddressList for ControlledAddressList.Data;` is saying that we should use the functions defined in the `ControlledAddressList` library and apply those functions on a custom datatype defined in `ControlledAddressList.Data`.

Summary

In this chapter, we looked at defining and creating a Solidity contract. We looked at creating custom modifiers for the functions, understood the different types of functions you can define inside a contract to architect your contract behavior, learned how to create and emit events for logging purposes, and dug deeper into Solidity inheritance by creating interfaces and abstract contracts. We also learned how to write custom user-defined libraries and use them in our contracts.

In terms of the Solidity language, we have covered all the topics that you are required to know about in order to write a basic Solidity contract. These topics cover almost 99% of the language's features, although there are ways to write assembly language blocks in Solidity as well. However, this is an advanced topic that you should take a look at in your own time. Assembly language usage should be a last resort, since using it increases the chances of your contract being prone to security issues.

We have covered the Solidity language while keeping two versions of it in mind: 0.4.25 and 0.5.0. However, in the future, the language may add, remove, or update some its features. You should always check which version of the Solidity language is the latest one, as well as if there are any bug fixes, so that you can make an informed choice about which version you install in the future.

In the next chapter, we will look at the tools that are used to create, test, and deploy contracts. We will also look at how to set up MetaMask and Remix IDE, as well as how to use them.

Questions

1. How big can a contract be in terms of code lines?
2. I have seen some contracts (deployed on `etherscan.io`) with their contract name being used for the function name, which is then used as the constructor of the contract. Why is this way of defining a constructor deprecated?
3. When should fallback functions be used?
4. Is using a `for` or a `while` loop recommended?
5. How do you perform a complex search or filter operation in Solidity?
6. Can you add custom functions for default data types that are supported in Solidity?
7. Why has the `throw` keyword been removed from the Solidity language?

Section 2: Deep Dive into Development Tools

There are different tools available to develop bug-free smart contracts in Solidity. The tools available for development, testing, and debugging are Truffle, Ganache, Remix, and MetaMask. We will cover code quality improvement tools, such as Solium, Solhint, and Surya, with a view to ascertaining possible bugs and issues. The reader will get to explore these tools.

The following chapters will be covered in this section:

4
Learning MetaMask and Remix

In previous chapters, we learned about blockchain and what a smart contract is. A smart contract should be used only when it is required to use the programmable properties of the blockchain. We looked into the most famous contract-oriented language, Solidity. We also understood the basic syntax of the language and how to write smart contracts in Solidity. As we have learned the language, we now need to learn about some tools that are useful for developing Solidity contracts.

In this chapter, we will look at tools that are used to create, test, and deploy contracts. We will look at how to set up MetaMask and the Remix IDE and how to use them.

The following topics will be covered in this chapter:

- Setting up the MetaMask browser plugin
- Understanding transfers, networks, and accounts on the MetaMask plugin
- Using the Remix IDE for contract development
- Testing contracts with the Remix IDE
- Setting up a local instance of the Remix IDE

Technical requirements

In this chapter, we are going to look into MetaMask and the Remix IDE for Solidity contract development. To install and configure those tools, you first need to satisfy their technical requirements:

- Latest version of npm and Node.js.
- **MetaMask**: MetaMask is an open source, internet browser plugin that you can install from `https://metamask.io/`. This is officially supported for Google Chrome, Firefox, Opera, and Brave internet browsers. There are no additional requirements for working with MetaMask. The MetaMask source code can be found on GitHub at `https://github.com/MetaMask/metamask-extension`.

- **Online Remix IDE**: The Remix IDE is an online version of the Solidity editor and **Integrated Development Environment (IDE)**. You can access the IDE by opening the `https://remix.ethereum.org` in your internet browser.

- `remixd`: `remixd` is an open source, command-line tool to open contract files present on your local machine in the online version of the Remix IDE. The `remixd` source code can be found on GitHub at `https://github.com/ethereum/remixd`.

- `remix-ide`: `remix-ide` is the open source, locally-installed version of Remix IDE. You first need to install `npm` and `node.js`. The Remix IDE source code can be found on GitHub at `https://github.com/ethereum/remix-ide`.

Using the MetaMask plugin

MetaMask is a plugin for internet browsers, which lets you connect to the Ethereum blockchain. You can connect and interact with distributed applications using MetaMask without installing a local full blockchain node. It provides the UI to sign the transactions and keep your identities for different distributed applications.

You can connect to the Ethereum blockchain by installing the Ethereum wallet software provided by the Ethereum Foundation itself. However, it requires the full Ethereum blockchain to be downloaded to your machine before initiating any transaction. Downloading the full blockchain ledger would require **gigabytes (GBs)** of hard-drive storage space and necessitates keeping the wallet software running all of the time. Hence, it is preferred to use the MetaMask plugin, which is fast, provides secure communication, and needs a negligible amount of hard drive storage.

The MetaMask plugin comes under the hot wallets category. Let's discuss these wallet categories.

There are three different categories of cryptocurrency wallets to store and hold your coins and tokens such as ether:

- **Hot wallets:** Hot wallets are used for quick access to your cryptocurrency and are readily available for use on your internet-connected device. The private keys of the wallets are stored on your device. MetaMask is an example of a hot wallet.

- **Cold wallets:** Cold wallets are used to keep your cryptocurrencies offline and are not easily accessible on your electronic device for transactions. Normally, cold wallets are used in the form of paper wallets, where you keep the private keys printed on a piece of paper and not on any internet-connected device.
- **Hardware wallets:** Hardware wallets are also a form of cold wallet. Hardware wallets store the private keys of the wallet in a special device built to keep your cryptocurrency secure and not connected to the internet. There are some hardware wallets available on the market for purchase, including Ledger, Trezor, KeepKey, and CoolWallet S.

Looking into the definition of the different wallet categories, we find that MetaMask falls under the category of hot wallets. However, it also supports connecting to the Ledger and Trezor hardware wallets. Hence, if you are building any decentralized application that requires hardware wallet support as well, you can use this integration via the MetaMask plugin.

You can install the MetaMask plugin on various internet browsers such as Chrome, Firefox, Opera, and Brave. In future, there might be support for some other internet browsers too.

Let's install and setup the MetaMask browser plugin in a supported browser.

Installing and setting up the MetaMask plugin

The trusted source for installing the MetaMask plugin is the MetaMask website, at `https://metamask.io/`. From there, you can get the link to the plugin page appropriate for your internet browser. Ensure that you do not install this plugin from any other non-trusted website.

Once you have installed the MetaMask plugin in your internet browser, you should be able to see an orange fox icon in your browser plugin bar. First, you need to set up your MetaMask instance to get it started. Let's learn how to set this up on your machine.

 The MetaMask screenshots may not be the same when the new version of MetaMask is released.

First, you should see a welcome screen:

The MetaMask welcome screen

This is then followed by the **Create Password** or **Import with seed phrase** screen:

Create a new account by entering password or import with existing seed phrase

If you don't have a seed phrase previously generated using MetaMask, you should create a new one; otherwise, you can import an existing one by clicking on **Import with seed phrase**. To create a new seed phrase, input the password that you would like to set to open MetaMask—this will be used every time you open it. After that, follow the steps and accept the terms and conditions of use.

After this, you will see the screen with your unique 12-word secret backup seed phrase, as follows:

Your secret backup phrase generated by MetaMask

This secret seed phrase is generated using a standard algorithm called **BIP-44**, also known as **Hierarchical Deterministic (HD)** wallets (**BIP** is short for **Bitcoin Improvement Proposals**). This seed phrase is also called a **mnemonic**. Write down this seed phrase on a piece of paper in the sequence it is mentioned. Keep this seed phrase at a secret location because if your system crashes, you can restore your funds and tokens using this seed phrase. MetaMask also gives you the option to download the seed phrase in a text file on the same screen.

If this secret seed phrase is lost, there is no way to recover your funds or wallets. These seed phrases are generated on your computer only using MetaMask; hence, it is not stored online. Also, MetaMask does not create any different account online for you. For security reasons, do not keep your seed phrase on any online notepad like service, for example, Evernote.

In the next screen, you will have to confirm your generated seed phrase in the order the words are generated, as shown:

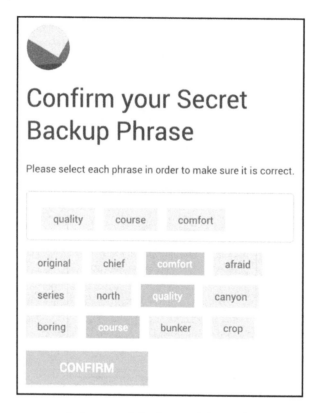

Confirm your secret backup phrase screen in the same order

Once your secret seed phrase is confirmed and the account is created, you will see your Ethereum wallet created, as shown:

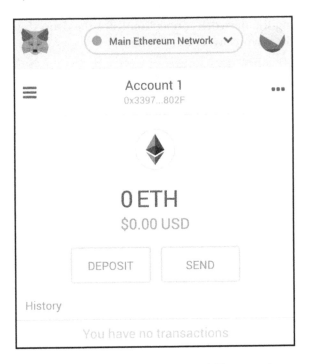

An EOA account is created and is connected to the main Ethereum network

As shown, you have **0 ETH** balance on the **Main Ethereum Network** as it is connected to the mainnet which needs real ether coin (which has economic value) for reflecting a non zero value. You can also add ERC20 standard tokens to the MetaMask wallet by using an ERC20 contract address.

Connecting to different Ethereum networks

You can connect your same EOA account to different Ethereum networks. You can change your network by selecting the network connection from the drop-down menu present at the top-center of the MetaMask window, as shown here:

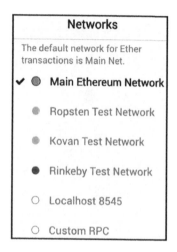

The list of Ethereum networks select from

As shown in the preceding screenshot, the list of networks available to connect to are as follows:

- **Mainnet**: Each ETH present on this network has a real world economic value
- **Testnet**: Each ETH has $0 real world economic value present on these networks:
    - **Ropsten Test Network**
    - **Kovan Test Network**
    - **Rinkeby Test Network**
- **Custom**: Each ETH has $0 real world economic value present on these networks:
    - **Localhost 8545**
    - **Custom RPC** connection URL

If you want to perform transactions on the mainnet, you should have real ETH and can do transactions by connecting to the **Main Ethereum Network**.

The localhost and custom RPC are there to connect to your local installed blockchain. You can configure the default ETH balance for your local installed blockchain. In the next chapter, we will look into how to connect to your Ganache instance (your local blockchain) using these options.

Getting test ether from faucets

Each of the testnets has their own faucets available to get the free ETH to your wallet. Here is the list of faucet links that you can use to request free ether:

Test network name	Faucet link
Ropsten	`https://faucet.ropsten.be/`
Kovan	`https://faucet.kovan.network/`
Rinkeby	`https://faucet.rinkeby.io/`

The test networks' faucet links

As these testnet ethers are limited and mined by someone else who is sharing their ether with these faucets, you will get a limited quantity of ether with these links. Nonetheless, even one ether would be sufficient for you to deploy multiple contracts. However, if you have installed your own local blockchain, you get as many as you want.

Other features of the MetaMask plugin

There are many other features of the MetaMask plugin that you can use:

- Send ether to another Ethereum EOA/contract using the **SEND** button
- Create new EOA on the same MetaMask
- Add any ERC20 tokens and receive them
- Import any EOA account with its private key
- Export an account's private key

Using the Remix Solidity IDE

To write Solidity contracts, the **Ethereum Foundation** has provided an online version of the Solidity IDE. You can open the Remix IDE in your internet browser by going to `https://remix.ethereum.org`. This online tool has the basic features required to write Solidity contracts and compile, test, and deploy your contracts.

 Ethereum Foundation is responsible to take decisions for the future development of the Ethereum blockchain.

As it's an online version, it is always updated with the latest features and compilers. You don't need to do any updates if you are using the online version. Let's learn how to use the Remix IDE.

The Remix IDE overview

Once you download and open the Remix Solidity IDE, you should see a screen like the following:

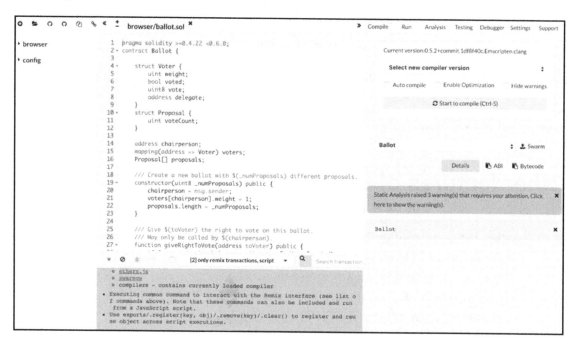

Remix Solidity IDE full-screen view

There are multiple sections:

- The left pane is the file browser, where you can choose or create new Solidity files
- The middle-top area is to write your Solidity code in the contract file
- On the middle-bottom area, you will be able to see details about the transactions
- The right pane is to show the different tools that you can use in the IDE, for example, to **Compile**, **Run**, and **debug** Solidity files

Compiler tools present under the Compile tab

As you can see on the right panel in the Remix IDE, the **Compile** tab is open. Here, you can choose which compiler version you would like to compile your contacts with. In this tab, there are three check boxes present, which are as follows:

- **Auto compile**: This is to auto compile your contract code as you keep on typing your contract in the contract editor area.
- **Enable Optimization**: The Solidity compiler supports code optimization to reduce the size and instructions present in the code. The behavior of the code will be the same even after optimization; however, the gas cost required to deploy the contract and make transactions would be greatly reduced. It is recommended to enable optimization when deploying your contract on the mainnet.
- **Hide warnings**: When you don't want to list compiler warnings in the right-bottom panel of the IDE, you can check this.

There are some other buttons, as well, which give more details about the contract:

- **Details Button**: When you click this, it will open a pop-up screen and show the different details about the contract, such as the **metadata**, **bytecode**, and **Abstract Binary Interface (ABI)**.
- **ABI**: You can click on this button to copy the contract ABI in JSON format to your clipboard.
- **Bytecode**: You can click on this button to copy the contract's bytecode in JSON format to your clipboard.

Just below these buttons, you can see the compilation errors and warnings present in the contract code.

Understanding the Run tab

The **Run** tab in the Remix IDE is mostly used for deploying the contract and accessing the functions of a contract. Using this tab, you can do the following operations:

- Choose the VM environment to connect with
- Deploy a new contract
- Attach the code to an already-deployed contract
- Change EOA accounts to use or initiate transactions with

- Set transaction parameters
- Access and call the `public` and `external` functions present in the contract
- Access `public` state variables

As shown in the following screenshot, the **Run** tab looks like this and has different subsections to use:

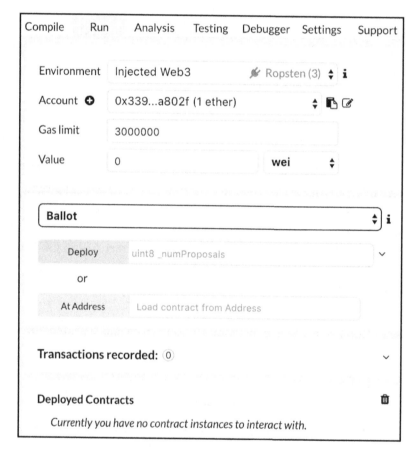

The Remix IDE Run tab

Let's understand each functionality of the **Run** tab in detail.

Selecting the environment to connect with

Using the **Environment** drop-down menu, you can choose the environment to connect your Remix instance. The Remix IDE supports connecting with three environments:

- **JavaScript VM:** The Remix IDE itself, to some extent, behaves like a blockchain. If your contracts are simple and can be run inside the browser, you would be able to deploy and test your contract using the VM provided by the browser itself. Using the JavaScript VM is fast and easy; however, it does not work with complex deployment.
- **Injected Web3:** Plugins such as MetaMask have integration with the Remix IDE. Using this environment, you can connect your MetaMask account with Remix. As you can see in the preceding screenshot, it is connected with **Injected Web3** (with MetaMask installed in the browser) and is using the **Ropsten** test network.
- **Web3 Provider:** When you have your local blockchain setup running on your local machine or on another machine, you would be able to connect to that blockchain using this environment. To connect using this, you would need to have a URL to connect to your blockchain.

Choosing different wallet accounts

You can choose different EOA account addresses to initiate transactions with. Under the **Account** drop-down list, it shows you the EOA accounts connected with Remix. As seen, the Remix instance is connected using MetaMask and an EOA that starts with 0x339; this also has a balance of 1 ether on the **Ropsten** test network. You can choose between multiple accounts that you want to use depending on your requirements.

If you are using the **Injected Web3** environment and are connected with MetaMask, this drop-down list will only show one account in the list–that is, the account you specify from MetaMask itself. Once you change the address on MetaMask, it will automatically be reflected in the Remix IDE and the account will be changed. However, if you are using the **JavaScript VM** environment, it would populate the drop-down list with some random account addresses; each will have 100 ether balance and you can choose different accounts from this list.

Transaction parameters

When you initiate any transaction from the Remix IDE, you will have to supply the values for the two transaction fields:

- **Gas limit:** When you are making any transaction using the Remix IDE, you need to set an appropriate gas limit for that transaction so that it can execute as expected. By default, it is set to 3 million units of gas, which is sufficient for many transactions. However, if your transaction requires special care and needs more gas units, then you can change this here before initiating your transactions.
- **Value:** This is the value of the ether that you want to send along with your transaction. This is required mostly if you are calling a `payable` function on a contract and if, for that function call, you want to transfer some ether as well. Along with the value, you can select the value unit, as well, from the drop-down list. Units are in wei, gwei, finney, and ether.

Selecting the contract to use

There is a drop-down list present in the **Run** tab that contains all the contract names that are linked to the contract that you have opened in the contract editor area. Once you have selected the contract name, you are allowed to deploy a new contract or connect to the existing contract.

This list shows only the contracts that are linked and accessible with the contract that you have opened on your contract editor. If you change the contract on your contract editor, this list will also change according to the contracts selected and linked.

Using deploy and attach

Just under the drop-down list of contract names, you have two buttons, **Deploy** in red and **At Address** in light blue, as shown in the preceding screenshot:

- The **Deploy** button: Using this button, you can deploy a whole new contract. The contract will be deployed on the selected environment and using the chosen EOA account. Every deployment of a contract generates a new address, on which the contract is deployed. This new address will be returned and your Remix IDE will, by default, be attached to that newly deployed contract. Hence, you can call functions of that contract just after that. There are some contracts that require parameters to be passed into their constructor to deploy. Just after the **Deploy** button, you will see a text box in which you can pass in the parameters to the contract's constructor, separated by , (a comma).

- The **At Address** button: Using this button, you can connect your contract code to an already deployed contract address. You must know at which address the contract is deployed on the network. Use that address and paste it into the text box next to this button and click on **At Address**. This will attach your contract code (opened in the contract editor) to the deployed contract. Then, you will be able to call functions and read the values of the contract.

Deploying a contract

We have a sample contract that we will deploy using the Remix IDE and MetaMask. We will interact with the contract and learn how to see the behaviors of the contract.

As shown in the following `DeploymentExample` code, we have a contract that takes ether as a deposit from an account, and the same account can withdraw its deposit as well. An account can deposit ether by calling the `depositEther()` function, or by just sending the ether to the deployed contract address itself. This will trigger the fallback function of the contract, which in turn calls the `depositEther()` function itself:

```solidity
pragma solidity ^0.5.2;

import "github.com/OpenZeppelin/openzeppelin-
solidity/contracts/ownership/Ownable.sol";
import "github.com/OpenZeppelin/openzeppelin-
solidity/contracts/math/SafeMath.sol";


contract DeploymentExample is Ownable {
    using SafeMath for uint;

    mapping(address => uint) public balances;
    uint public totalDeposited;

    event Deposited(address indexed who, uint amount);
    event Withdrawn(address indexed who, uint amount);

    function() external payable {
        depositEther();
    }

    function depositEther() public payable {
        require(msg.value > 0);
        balances[msg.sender] = balances[msg.sender].add(msg.value);
        totalDeposited = totalDeposited.add(msg.value);
        emit Deposited(msg.sender, msg.value);
```

```
        }

        function withdraw(uint _amount) public {
            require(balances[msg.sender] >= _amount);
            balances[msg.sender] = balances[msg.sender].sub(_amount);
            totalDeposited = totalDeposited.sub(_amount);
            msg.sender.transfer(_amount);
            emit Withdrawn(msg.sender, _amount);
        }

        function kill() public onlyOwner {
            selfdestruct(address(uint160(owner())));
        }
    }
```

To deploy the preceding contract, copy paste the preceding code in the code editor and then compile the contract. You can also follow the steps provided in the following section of this chapter to use `remixd` to open your contracts in the Remix IDE. Once the contract is compiled successfully, it should show the following option on the right-hand side panel of the editor. There, you have a red button to deploy the contract:

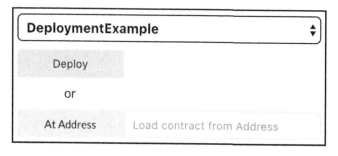

Contract deployment panel

For this deployment, we are using the **Injected Web3** environment and are connected with MetaMask. We are using the **Rinkeby Test Network**. Once you click on the **Deploy** button, MetaMask will open a popup and ask for confirmation to deploy the contract, as shown in the following screenshot:

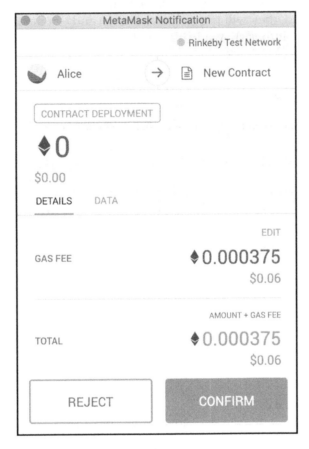

MetaMask contract deployment confirmation screen

Once you click the **CONFIRM** button shown in the MetaMask popup screen, it will initiate a transaction to deploy a new instance of the contract. Once it is deployed successfully, the Remix IDE will show method-specific buttons on the right panel. There, you see all of the public functions and public state variables; you can access those functions and initiate transactions. As shown in the following screenshot, the buttons shown in red are the functions; when initiated, they will trigger a transaction and can modify the state of the contract. However, the buttons shown in blue will initiate calls to the contract and will not make state changes to the contract:

Functions allowed to call on a deployed contract

As you can see in the preceding screenshot, the address of the deployed contract is `0xDAc483f4441159Cb773F7421495a458e32707500` on the Rinkeby test network.

Initiating a transaction to execute the function

As shown, we have deployed the contract using **Alice's** Ethereum account, present in MetaMask. Just by deploying the `DeploymentExample` contract, Alice became the owner of the contract and only Alice's Ethereum account can call the `kill()` function of the contract. Other Ethereum accounts are not allowed to call this function.

Let's use **Bob's** Ethereum account and initiate some transactions on the contract. We will call the fallback function of the contract. There are two ways to initiate the transaction to the fallback function—you can either use the Remix IDE and call the fallback function, or you can directly send the ether to the contract address, `0xDAc483f4441159Cb773F7421495a458e32707500`, deployed on the Rinkeby test network.

As shown in the following screenshot, we are sending **0.5 ETH** to the contract from Bob's account directly (without the Remix IDE):

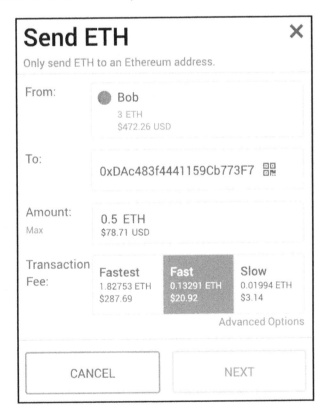

Bob sending 0.5 ether to the contract address

As you can see in the preceding screenshot, that user can choose between different **Transaction Fee** options. The fastest transaction execution would require a higher transaction fee in ether; on the other hand, a slower transaction would require less of a fee in ether. You can also change those values as per your custom settings; you can do so by clicking on the **Advanced Options** link. Once the 0.5 ether is transferred to the contract, you will be able to see the 0.5 ether balance on contract account on the block explorer (rinkeby.etherscan.io) as well.

Let's make another transaction and call the depositEther() function via the Remix IDE. We will initiate this transaction from **Charlie's** Ethereum account. To do this, change account in MetaMask. As shown in the following screenshot, we have the value as 0.3 and selected the unit as **ether**. This will send the **0.3 ether** with the transaction:

Set the value to send ether with the transaction

Now, click on the red depositEther button to initiate the transaction to the function. This will open the MetaMask confirmation screen to confirm the transaction. Once the transaction is confirmed, it will send 0.3 ether and execute the depositEther() function of the contract.

Initiating a call to a view function and state variables

As we have deployed and initiated some transactions in the last section, let's access the view functions of this deployed contract.

When accessing view functions or state variables, you do not need to initiate the transaction and the MetaMask confirmation is not required. These are known as **message calls**. All of the message calls to a contract are free and can be executed as many times as you require. The message calls do not make any modification on the blockchain data.

You can click on the public view functions, and public state variables are shown in blue:

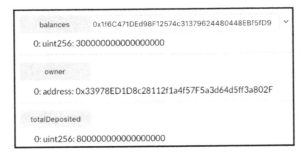

balances	0x1f6C471DEd98F12574c31379624480448EBf5fD9	⌄
0: uint256: 300000000000000000		
owner		
0: address: 0x33978ED1D8c28112f1a4f57F5a3d64d5ff3a802F		
totalDeposited		
0: uint256: 800000000000000000		

Accessing view functions and state variables

As shown in the preceding screenshot, the `totalDeposited` state variable returns **0.8 ether** (shown in wei in Remix); the `owner` address is shown; and the balance of an account address is **0.3 ether** (shown in wei).

Connecting the contract folder using remixd

`remixd` is a tool for the Remix IDE only. This tool opens a WebSocket communication between `remixd` and the Remix IDE. Using this tool, you can provide a folder to be shared with the Remix IDE. This tool is useful when you have multiple contracts created on your local machine and you would like to use the Remix IDE to deploy or attach to a contact. As putting each contract into the Remix IDE would be a cumbersome job, to reduce that pain for the developer, the `remixd` tool is used.

You can install the `remixd` tool on your machine using the following command:

```
$ npm install -g remixd
```

Once it is installed, you can use the `remixd` command to share the folder with the Remix IDE. To share the folder, you have to execute the following command:

```
$ remixd -s <absolute-path> --remix-ide https://remix.ethereum.org
```

The preceding code is the syntax to use for the `remixd` command. Once you have the `MasteringSolidity` code obtained from GitHub, first, change directory to the `Mastering-Blockchain-Programming-with-Solidity/Chapter04/` folder and run the following command:

```
$ remixd -s . --remix-ide https://remix.ethereum.org
```

If your `remixd` command is successful, you will see the following log output:

```
Thu Dec 27 2018 02:49:43 GMT+0530 (IST) Remixd is listening on
127.0.0.1:65520
```

Ensure that you keep the instance running so that the connection to the Remix IDE remains established.

Using the preceding command, we are sharing the full `MasteringSolidity/Chapter04` folder, as we also need our `node_modules` folder to be shared. Because we are referring to some Solidity libraries, those are installed under the `node_modules` folder. We are using `openzeppelin` library contracts, those are installed and present under the `node_modules/openzeppelin-solidity` folder. The following are the code lines from the contract that `import` the contracts from OpenZeppelin:

```
import "openzeppelin-solidity/contracts/ownership/Ownable.sol";
import "openzeppelin-solidity/contracts/math/SafeMath.sol";
```

Now, as you have shared the `MasteringSolidity/Chapter04` folder, you should be able to connect your folder using the Remix IDE. To connect, follow these steps:

1. Open your internet browser and go to `https://remix.ethereum.org`, which will open the online version of the Remix IDE.

2. At the top-left corner, there is a link button, as highlighted in the following screenshot.

3. It will open a pop-up box to ask permission to connect. Click on **connect**.

4. Once it is connected, you should see the `localhost` folder added to the contract explorer pane present on the left:

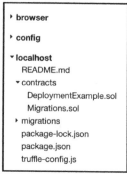

The left-hand contract explorer panel shows localhost once remixd is connected

5. You should be able to select and open any of your contracts and it will link the dependencies of the contract automatically:

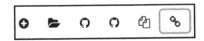

Link button to connect to remixd

Setting up a local instance of the Remix IDE

We have used the online version of the Remix IDE so far in our example. However, to use the online version, you will always need an internet connection and if you would want to work offline, it will create problems for you. Hence, you can also install and use a local version of the Remix IDE. To install the Remix IDE locally, follow the commands:

```
$ npm install remix-ide -g
$ remix-ide
```

 To install `remix-ide` on Ubuntu, use the following command: `sudo npm install remix-ide -g --unsafe-perm=true --allow-root`

Using the blockchain explorer at etherscan.io

You can see the status of the transaction, and the state of the wallet and contract accounts on `etherscan.io` (the Ethereum block explorer). As we have deployed the `DeploymentExample` contract in this chapter, we can see the status of the contract by opening `https://rinkeby.etherscan.io` and searching for the contract address.

As shown in the following screenshot, the status of the contract on etherscan is that it has a balance of **0.8 ether**. The contract code is also published for this contract; you can see the contract code by opening the **Code** tab.

During the transaction, different events the contract has triggered can be seen also under the **Events** tab:

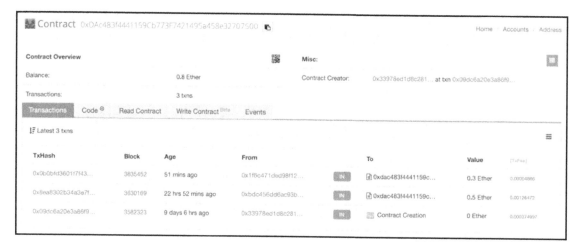

The contract status on the block explorer, etherscan.io

Using etherscan, you can also access all of the public view functions and public state variables. To access these, open the **Read Contract** tab and it will list them.

On the block explorer, you can also see the details regarding each transaction and block. Block explorers are the easiest and fastest way to access information about the blockchain.

Ethereum wallet software

You can download Ethereum wallet software and access your wallet account and deploy contract accounts. However, to perform these operations, you would have to download the blockchain. This makes Ethereum wallet software less developer-friendly. On the other hand, the MetaMask, Remix, Truffle, and Ganache tools provide fast and developer-friendly environments. Hence, Ethereum wallet is not used mostly by the developers.

There might be some special requirements when you are deploying your own private Ethereum network; in that case, Ethereum wallet should be used.

We are not covering Ethereum wallet's features in this book. The reader can download and play with it to understand its features and limitations. You can download Ethereum wallet software Mist from https://github.com/ethereum/mist/releases.

Parity also provides some Ethereum software called Parity Ethereum. You can download it from `https://www.parity.io/ethereum/`.

Using myetherwallet.com

There is a website called `https://www.myetherwallet.com/`, which also provides tools to access the Ethereum blockchain and initiate some special transactions:

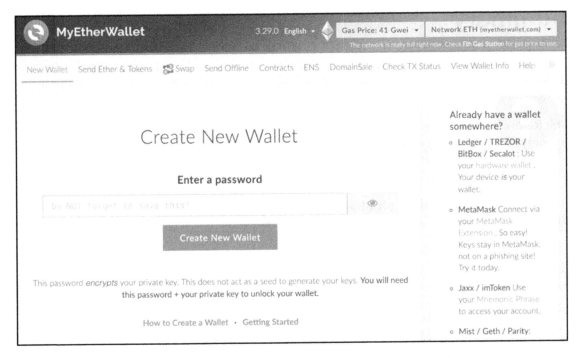

The myetherwallet.com website provides tools for Ethereum

As shown in the preceding screenshot, using this, you can perform the following operations:

- Create a new Ethereum paper wallet. When the private key is printed on paper and kept secret this is called a **paper wallet**.
- Connect using different hardware wallets and different wallet formats and initiate the transactions.

- Just like the Remix IDE can access contract functions, in the same way, you can also access these functions via `myetherwallet.com` using the ABI of the contract only.
- Using **Send Offline**, you can sign the Ethereum transaction from an offline (air gapped) computer to improve security.
- Using **Ethereum Name Service** (**ENS**), you can register a human-readable name for your wallet address.

Summary

We learned about MetaMask and the Remix IDE in this chapter. As of the time of writing, MetaMask is the easiest and most user-friendly way to access **Decentralized Applications** (**DApps**). It is also developer-friendly for building Ethereum contracts and interacting with the Remix IDE.

The Remix IDE provides a great way for developers to deploy and access contracts. It was the only way for the developers to do so before the Truffle framework was developed. However, after using the Truffle framework, developers may use Remix IDE to access already-deployed contracts and call functions.

It is difficult to test and debug contracts on the Remix IDE. The Truffle framework provides tools for this, which we will look into in the next chapter.

Questions

1. How many accounts can I create on MetaMask?
2. Can I connect MetaMask to my own private blockchain?
3. What is the MetaMask secret seed and when would it be used?
4. Can MetaMask receive ERC20 tokens?
5. Can I cancel a pending transaction using MetaMask?
6. How can I access the JSON ABI for the contract using the Remix IDE?
7. Should I always use optimization for contracts?

5
Using Ganache and the Truffle Framework

In the last chapter, we learned about using the MetaMask browser plugin and the Remix IDE to write and deploy contracts. These tools are good for small projects and a smaller number of contract files. However, when you are working on a large project and have multiple contract files to manage, these tools are not developer-friendly.

Using Ganache, you can deploy your own local blockchain on your machine; this is used for testing your contracts. This is way faster than using a testnet, and is even faster than Kovan, which is the fastest testnet out there.

Truffle framework, along with Ganache, provides all of the tools that are needed for developers to work on many contract files. You can write test cases, migration scripts, and debug your contract using the framework.

In this chapter, you will learn to use Ganache blockchain along with Truffle framework commands. There are different commands present in Truffle framework to compile/test/debug/migrate your contracts files. Also, you will learn how to write contract migration JavaScript files.

We will cover the following topics in this chapter:

- Installing and using Ganache blockchain
- Advanced settings in Ganache for custom needs
- Installing and configuring Truffle framework
- Writing contract migration scripts
- Running the test suite
- Debugging contract transactions using the Truffle debug console

Technical requirements

The Solidity contracts, tests, and migration scripts used in this chapter are present on GitHub at `https://github.com/PacktPublishing/Mastering-Blockchain-Programming-with-Solidity/tree/master/Chapter05`.

In this chapter, we are going to learn how to use Ganache to set up our local blockchain, and Truffle framework for the contract development environment. There are some pre-requisite tools that should be installed before installing Ganache and Truffle.

Ganache (local blockchain): You can use either of the Ganache to run your local blockchain:

- **Ganache GUI**: To install the GUI version of Ganache tool, you should have `npm` version v5.3.0 or later and NodeJS version v8.3.0 or later installed. The source code of the project can be found on GitHub at `https://github.com/trufflesuite/ganache`.
- **Ganache CLI**: To install the open source **Command Line Interface (CLI)** version of Ganache, you should have NodeJS version 6.11.5 or later installed. The source code of the project can be found on GitHub at `https://github.com/trufflesuite/ganache-cli`.

Truffle: To install and use the `truffle` command, you require the following tools:

- You need NodeJS v8.9.4 or later.
- Supported on Windows, Linux, or macOS X.
- It also needs a blockchain to connect, which should support the JSON RPC API. There are different flavors of Ethereum blockchain that you can use with Truffle.
- You can use any code editor to create and edit your Solidity contract files, although there are some editors that have Solidity language support, such as Sublime, Eclipse, and IntelliJ IDEA.

Local blockchain with Ganache

Ganache provides the GUI-based local Ethereum blockchain development environment to deploy and test contracts. Once installed on the machine, Ganache provides single-click, up-and-running blockchain deployment. This is especially built to reduce the dependency on Ethereum wallet software, which requires a lot of resources to run the local blockchain instance. Using Ganache, you can start the local blockchain within seconds, and it does not require heavy CPU/memory or disk resources.

Ganache does not require any faucet links, as these are only required for Ethereum test networks. As it's a local blockchain instance, it provides a number of pre-initialized accounts and each of the accounts has test ether balances. Testing on Ganache is faster, at least 10x faster than the Ethereum test network, as Ganache processes the transactions instantly as it receives them.

The Ganache tool is the part of the Truffle software suite. You can download and install the latest version of Ganache from `https://truffleframework.com/ganache`.

Once you install Ganache on your machine, you can start it and should see the Ganache GUI, as shown in the following screenshot:

The Ganache welcome screen

As you can see, you can start the instance of a local blockchain just by clicking on the **QUICKSTART** button. You can also create a workspace with the **NEW WORKSPACE** button, and use some advanced features of Ganache.

Let's learn how to start your personal instance of the Ganache blockchain and configure different workspaces on it.

Starting a local blockchain

As mentioned earlier, you can start a local Ganache blockchain instance by just clicking on the **QUICKSTART** button. This starts the blockchain with a configured mnemonic phrase and initializes accounts by default with **100.00 ETH** each, as shown in the following screenshot. By default, it opens the **Accounts** view:

QUICKSTART Ganache blockchain running

You can go through the different view options available in Ganache using the following top bar:

The Ganache top bar to switch to different views

As shown in the preceding screenshot, the top bar provides the following view options to choose from. Let's understand when these options can be used and what kind of information they show:

- **ACCOUNTS:** Lists all accounts and their ETH balances, along with the number of transactions initiated from those accounts.
- **BLOCKS:** Lists the number of blocks generated and gas used per block. This is similar to the online block explorer (`http://etherscan.io`), where each block generated is shown along with the transactions included in it.
- **TRANSACTIONS:** Lists the transactions generated so far, along with their transaction ID, the gas used, and other details.
- **CONTRACTS:** Lists of all contracts from the Truffle project and their contract addresses, along with their deployment statuses.
- **EVENTS:** Lists all of the events triggered during transaction execution.
- **LOGS:** Opens the console log of Ganache.

You can also see the status of the Ganache GUI on the status bar, as shown in the following screenshot:

Status of the Ganache blockchain

As shown in the preceding screenshot, the status bar of Ganache shows the following information:

- **CURRENT BLOCK:** The number of blocks generated so far
- **GAS PRICE:** The gas price per transaction in wei
- **GAS LIMIT:** The gas limit per block
- **NETWORK ID:** The network ID of the blockchain
- **RPC SERVER:** The blockchain by default starts an RPC server on `http://127.0.0.1` (localhost) and at port `7545`
- **MINING STATUS:** There are two types of mining—**AUTOMINING** means the transactions will be processed instantly when they are received at Ganache
- **WORKSPACE:** The workspace name being used

Creating workspaces for projects

In the new release of Ganache v2.0.0, a new workspace feature was added. Using this feature, you can configure your Truffle projects into a workspace and Ganache would be able to show details specific to the contracts and events triggered:

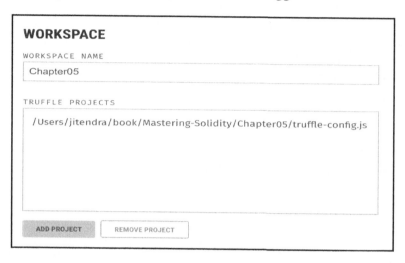

Ganache workspace creation

As shown in the preceding screenshot, we have configured the `MasteringSolidity/Chapter05` Truffle project into Ganache. You can add more Truffle config files into the same workspace. Ganache will scan these projects and list all of the Solidity contracts found into the contracts view present in Ganache.

Once you start Ganache with a workspace instance, it will also save the blockchain state. This will allow you to restore to the old state of blockchain when the workspace is restarted next.

Ganache advance configuration

You can change the default configuration of Ganache via the settings. You can see the following configuration bar in the settings GUI, as shown in the following screenshot:

Ganache advanced settings

As shown in the preceding screenshot of the advanced settings bar, in Ganache, you can configure the following:

- **WORKSPACE:** Set up the project workspace and choose the `truffle.js` or `truffle-config.js` file to link contract files
- **SERVER:** You can change the default server address and port number to start the Ganache server
- **ACCOUNT & KEYS:** Configure the number of default EOA accounts to be created as well as the ETH balance for each EOA account
- **CHAIN:** You can change the gas limit (to be used per block and per transaction) and gas price per transaction
- **ADVANCED:** You can change the log settings to generate more verbose logs, required when debugging transactions using Truffle

The command-line version of Ganache

Just like Ganache GUI version, it also comes with a CLI version called `ganache-cli`. The command-line version is useful in writing automation scripts. This also provides all the options available in the GUI version of Ganache.

To install `ganache-cli`, run the following command:

```
$ npm install -g ganache-cli
```

The preceding command will install `ganache-cli` globally, and the command will be accessible from anywhere.

To start the Ganache with the default configurations, you can run the following command:

```
$ ganache-cli
```

The command will start the Ganache and output the following logs:

```
Ganache CLI v6.2.5 (ganache-core: 2.3.3)

Available Accounts
==================
(0) 0x1abd8d35a002b1b1cf19b9fc5cea9c914d08d72c (~100 ETH)
(1) 0xbbdb8ca508829ce18de752e04ca96e3da6482795 (~100 ETH)
(2) 0xb07b2bffbb993937851006101c76d8bcae06543e (~100 ETH)
(3) 0x7bc6ff3576dd6a9df6a9c32b92af555d4ccbdea8 (~100 ETH)
(4) 0x87dabdef8742cee59df58db38698cd1117e208af (~100 ETH)
(5) 0x330ab1e8cae9a1f895629c39d28e961d2ca5a706 (~100 ETH)
```

```
(6)  0xdf77c139530245f4c9acd347e66061ad29e2b1df  (~100 ETH)
(7)  0x240ca8a5c486e80947a6f7d880a596d044954d2d  (~100 ETH)
(8)  0xb9b9b077c3abbf47eb07b307c999abae516c49ec  (~100 ETH)
(9)  0x41c6316d15bd3359c526a38435a20d8459ec61f7  (~100 ETH)

Private Keys
==================
(0)  0x6c6362718e3a1fa5bb2cd2a240f0bf8b34eac0a21594877590a6d68c9da64fe5
(1)  0x287c6678a50c17507409aae12112e481a5c7547c8bf5e993ec034f4eb392cf6f
(2)  0x008753b01bd98b585362ace7b12640457a5380ddfe0b20bc252a6819636b8b4d
(3)  0x9560397af5725d23c2bf418230c898883f67cc2fa42add0583fa72ad7866123c
(4)  0x5349625d1d37cc7c0cc2d86eded9458d340527c016351e247e74adbebdaeaa1a
(5)  0xd5a1354d939c3bb41ab7e2aaea0d6ce32edd887a2deccdd28b0929b2df71280b
(6)  0x9306c8e37db5f6c513b894d8d7416d5c439e2dd44154e1d6e867749a108e3709
(7)  0xd26c198c7ccab104315c43aa143ff3ed556b8c31a674cfa69e43a0bc16ff248a
(8)  0xebe8ba9cc4563a780be3271a05b21a1e144e58556f823709776854b7ad67d9a4
(9)  0x5213628c32d0e92c0a9f5d9df3c60f9e0d7d401ed31634cce318838ef66c3cfa

HD Wallet
==================
Mnemonic: country adjust orphan casual armed negative sorry found angle
medal narrow skate
Base HD Path: m/44'/60'/0'/0/{account_index}

Gas Price
==================
20000000000

Gas Limit
==================
6721975

Listening on 127.0.0.1:8545
```

As you can see in the above output, a Ganache instance is started and listening on port 8545 of your local machine. You can provide the IP address and port in your Truffle config file to use this Ganache instance. Also, the command has initialized 10 EOA accounts by default, and each account has **100 ETH** balance to use.

There are lots of options available for `ganache-cli`, you can get the help for these options by running the following command:

```
$ ganache-cli --help
```

The preceding command will display all available options as follows:

```
✔ ganache-cli --help
Network:
  -p, --port            Port number to listen on                              [number] [default: 8545]
  -h, --host, --hostname Hostname to listen on                            [string] [default: "127.0.0.1"]
  --keepAliveTimeout    The number of milliseconds of inactivity a server needs to wait for additional incoming data,
                        after it has finished writing the last response, before a socket will be destroyed.
                                                                              [number] [default: 5000]

Accounts:
  -a, --accounts            Number of accounts to generate at startup           [number] [default: 10]
  -e, --defaultBalanceEther Amount of ether to assign each test account         [number] [default: 100]
  --account                 Account data in the form '<private_key>,<initial_balance>', can be specified multiple
                            times. Note that private keys are 64 characters long and must be entered as an 0x-prefixed
                            hex string. Balance can either be input as an integer, or as a 0x-prefixed hex string with
                            either form specifying the initial balance in wei.                         [array]
  --acctKeys                saves generated accounts and private keys as JSON object in specified file
                                                                              [string] [default: null]
  -n, --secure              Lock available accounts by default (good for third party transaction signing)
                                                                              [boolean] [default: false]
  -u, --unlock              Comma-separated list of accounts or indices to unlock                      [array]

Chain:
  -f, --fork                Fork from another currently running Ethereum client at a given block. Input should be
                            the HTTP location and port of the other client, e.g. 'http://localhost:8545' or
                            optionally provide a block number 'http://localhost:8545@1599200'
                                                                              [string] [default: false]
  --db                      Directory of chain database; creates one if it doesn't exist [string] [default: null]
  -s, --seed                Arbitrary data to generate the HD wallet mnemonic to be used
                                                         [string] [default: Random value, unless -d is specified]
  -d, --deterministic       Generate deterministic addresses based on a pre-defined mnemonic.        [boolean]
  -m, --mnemonic            bip39 mnemonic phrase for generating a PRNG seed, which is in turn used for hierarchical
                            deterministic (HD) account generation                                    [string]
  --noVMErrorsOnRPCResponse Do not transmit transaction failures as RPC errors. Enable this flag for error reporting
                            behaviour which is compatible with other clients such as geth and Parity.
                                                                              [boolean] [default: false]
  -b, --blockTime           Block time in seconds for automatic mining. Will instantly mine a new block for every
                            transaction if option omitted. Avoid using unless your test cases require a specific
                            mining interval.                                                          [number]
  -i, --networkId           The Network ID ganache-cli will use to identify itself.
                            [number] [default: System time at process start or Network ID of forked blockchain if configured.]
  -g, --gasPrice            The price of gas in wei                          [number] [default: 20000000000]
  -l, --gasLimit            The block gas limit in wei                       [number] [default: 6721975]
  --allowUnlimitedContractSize Allows unlimited contract sizes while debugging. By enabling this flag, the check within
                            the EVM for contract size limit of 24KB (see EIP-170) is bypassed. Enabling this flag
                            *will* cause ganache-cli to behave differently than production environments.
                                                                              [boolean] [default: false]
  -t, --time                Date (ISO 8601) that the first block should start. Use this feature, along with the
                            evm_increaseTime method to test time-dependent code.                      [string]

Other:
  --debug        Output VM opcodes for debugging                              [boolean] [default: false]
  -v, --verbose  Log all requests and responses to stdout                    [boolean] [default: false]
  --mem          Only show memory output, not tx history                     [boolean] [default: false]
  -q, --quiet    Run ganache quietly (no logs)                               [boolean] [default: false]

Options:
  --help, -?  Show help                                                                            [boolean]
  --version   Show version number                                                                  [boolean]
```

Ganache CLI available options

You can use the `ganache-cli` command with the aforementioned options according to your needs.

As we have learned how to set up local instance of blockchain using Ganache, now let's learn to use Truffle framework.

Understanding Truffle framework

The Truffle suite offers `truffle` command-line tools to perform developer-centric tasks for Ethereum smart-contract projects. The `truffle` development environment provides network management, a testing framework, deployment, and a migration framework.

Using deployment and migration scripts, you can write scripts to deploy contracts and link them together. Using network management commands, you can manage deployment and migration states per network.

To install `truffle`, you need to execute the following command:

```
$ npm install -g truffle
```

As of writing this chapter, the latest version released by `truffle` is version v5.0.1.

As shown in the following screenshot, the following are the `truffle` command options available:

```
Jitendras-MacBook-Pro:MasteringSolidity jitendra$ truffle help
Truffle v5.0.1 - a development framework for Ethereum

Usage: truffle <command> [options]

Commands:
  build     Execute build pipeline (if configuration present)
  compile   Compile contract source files
  console   Run a console with contract abstractions and commands available
  create    Helper to create new contracts, migrations and tests
  debug     Interactively debug any transaction on the blockchain (experimental)
  deploy    (alias for migrate)
  develop   Open a console with a local development blockchain
  exec      Execute a JS module within this Truffle environment
  help      List all commands or provide information about a specific command
  init      Initialize new and empty Ethereum project
  install   Install a package from the Ethereum Package Registry
  migrate   Run migrations to deploy contracts
  networks  Show addresses for deployed contracts on each network
  opcode    Print the compiled opcodes for a given contract
  publish   Publish a package to the Ethereum Package Registry
  run       Run a third-party command
  test      Run JavaScript and Solidity tests
  unbox     Download a Truffle Box, a pre-built Truffle project
  version   Show version number and exit
  config    Set user-level configuration options
  watch     Watch filesystem for changes and rebuild the project automatically

See more at http://truffleframework.com/docs
```

truffle command options available

Out of the preceding commands, the following `truffle` command options are mostly used:

- `truffle init`: This initializes the new Truffle project and creates the required project files in an empty folder
- `truffle compile`: This command reads the Solidity compiler version, and other settings from the configuration file, loads the desired compiler version and compiles the contracts
- `truffle migrate`: This is the command to run the migrations scripts and deploy the contract on a given network configuration
- `truffle test`: This command runs the test suite using a given network configuration
- `truffle networks`: This command shows the different networks list and the deployed contracts on each network
- `truffle debug`: This command takes `transaction-id` and runs the debugger mode to debug the contract step by step

Setting up a Truffle project

You can initialize the Truffle project with the `truffle` default settings. To create a `truffle` project, run the following `truffle` command in an empty folder:

```
$ truffle init
```

The preceding command creates the following folder structure in an empty folder:

```
├── contracts              (Folder to keep .sol contract files)
│   └── Migrations.sol      (Default contract for migration)
├── migrations             (Folder to keep migration scripts)
│   └── 1_initial_migration.js (script to deploy Migrations.sol contract)
├── test                   (Folder to keep contract test cases)
└── truffle-config.js      (Default truffle configuration file)
```

This will create the following things in the folder to make it compatible with the Truffle framework:

- It creates the `contracts` folder, where you will keep all of your Solidity contract files.

- It creates the `contracts/Migrations.sol` file, which is required by the `truffle` project to keep track of the migration status of the contracts in each network. You should not make any changes in this file.
- It creates the `migrations` folder, where you keep all migration scripts of the project.
- It creates the `migrations/1_initial_migration.js` file to deploy the `Migrations.sol` contract. You should not make any changes in this file.
- It creates a `test` folder, where you will keep all of the test cases required to test the Solidity contract files.
- It creates a new `truffle-config.js` file, which keeps configurations of the `truffle` project, network, compiler.

Truffle configuration files

We are using Truffle v5.0.2 version for our project. You can see the full file contents of the `truffle-config.js` file present in the `MasteringSolidity/Chapter05` folder. We will look into different configuration settings you can perform using this file.

Configuring the Solidity compiler

In the config file, you will find the following settings for the Solidity compiler:

```
compilers: {
  solc: {
    version: "0.5.2",
    docker: false,
    settings: {
      optimizer: {
        enabled: true,
        runs: 200
      },
      evmVersion: "byzantium"
    }
  }
}
```

Let's look at each setting:

- `solc.version`: This takes the Solidity compiler version as a string parameter. You can change the Solidity compiler version according to your project requirements.

- `solc.docker`: Truffle supports fetching the Solidity compiler from a Docker image. If you have set this property as **true**, you must have Docker installed on your machine and already have fetched the `solc` (the command-line Solidity compiler) Docker image for that version. Truffle does not fetch the Docker image automatically. This property takes a Boolean parameter.

- `optimizer.enabled`: Optimization of the contract instruction will be performed at the time of compilation. This property takes a Boolean parameter.

- `optimizer.runs`: This defines for how many number of transactions you need your contract to be optimized. This property takes an integer parameter. Ideally, the value used for this property is `200`.

- `evmVersion`: This is an advanced configuration. This signifies the version of the EVM to compile the contracts for. This affects type checking and code generation. This property takes a string parameter. The following are the possible values for this property given to EVM while doing a soft-fork or a hard-fork of the Ethereum blockchain:

  - `byzantium` (the default)
  - `homestead`
  - `tangerineWhistle`
  - `spuriousDragon`
  - `constantinople`
  - `petersburg`

The optimized contracts sometimes take less gas cost at the time of deployment and function calls. It is recommended to always enable optimization.

Hard-fork is a process when a blockchain is getting upgraded with new features that are not backward compatible. However, a soft-fork is the process when the blockchain is upgraded with new features which are backward compatible. In Ethereum, each soft-fork or hard-fork is given a unique name to identify for example, `homestead`, `constantinople`, and others

Configuring networks

Under the network's properties, you can define multiple networks and their respective settings. These are the networks you can use for the testing and deployment of the contracts. The network setting also defines transaction parameters such as gas and gasPrice, which will be used while running the migration scripts. truffle also records the migration for each network and saves its state for the next run.

You can specify the network's name to use while running the migration:

```
$ truffle migrate --network rinkeby
```

Also, you can see each network's migration status, the contracts deployed, and the addresses on which they are present by running the following command:

```
$ truffle networks
```

The following is what the network setting looks like in the truffle-config.js file:

```
networks: {
  development: {
    host: "127.0.0.1",
    port: 8545,
    network_id: "*",
  },
  rinkeby: {
    provider: () =>
      new HDWalletProvider(mnemonic,
        `https://rinkeby.infura.io/<infuraKey>`),
    network_id: 4,
    gas: 5500000,
    confirmations: 2,
    timeoutBlocks: 200,
    skipDryRun: true
  }
}
```

In the preceding configuration example, we have defined two network configurations, development and rinkeby. The development and rinkeby are the names of the network; you can define any name you like for the network name. You can see their configuration parameters—the development network represents a blockchain running on a local machine at port 8545. And the rinkeby network uses HDWalletProvider to connect to the rinkeby test network using the Infura links.

Let's look at each property:

- `host`: This is the IP address of the host.
- `port`: This is the port number of the node open on the host.
- `network_id`: This is the network ID to match with. There are some predefined network IDs. You can use * (asterisk) to match any network ID.
- `provider`: The default web3 provider using `host` and `port` is used as `new Web3.providers.HttpProvider("http://<host>:<port>")`. If defined, the custom provider will be used; for example, `HDWalletProvider`.
- `gas`: This is the gas limit used for contract deployment. The default value is `4712388`.
- `gasPrice`: This is the gas price used for contract deployment. The default value is 100,000,000,000 wei (100 Shannon).
- `confirmations`: This is the number of block confirmations to wait between deployments. The default value is 0.
- `timeoutBlocks`: If a transaction is not mined, it keeps waiting for the specified number of blocks (the default is 50).
- `websockets`: You will need this enabled to use the `confirmations` listener. The default value is `false`.
- `skipDryRun`: This is used if you want to skip the dry run of the migration scripts on the network. The default value is `false`. This is useful for public networks, and allows you to dry-run the migration locally before performing the actual migrations on a public network.

You can get your Infura key by registering on `https://infura.io/`. This is a free service to generate keys for your access to the mainnet and different test networks.

For each network configuration, you should either define `host`/`port` or `provider`, but not both. There are different providers used internally.

Choosing which blockchain client to use

The Truffle framework requires the blockchain instance to run the test, deployment, or migration scripts on. There are different flavors of the blockchain instance that you can use. In this chapter, we have talked about Ganache, which is good for installing your local Ethereum blockchain and used for testing your Solidity contracts. Let's look at the different Ethereum blockchain software available that you can use, according to your needs.

Personal blockchain client

These blockchain instances are installed locally on your machine and used for testing purposes:

- **The Ganache GUI**: This is the GUI version of the Ganache tool, as we learned in this chapter. You should use this blockchain when you need to quickly look at the different accounts, transactions, blocks, contracts, and events being generated during the execution of the migration or test cases. You can also look at the gas costs consumed by each transaction in detail. You can also fine-tune the mining of the blocks using the advanced settings.
- **The Ganache CLI**: This is the CLI version of Ganache. This should be used when you need to automate your contract-testing environment. This is also useful in preparing your continuous integration environment. The transactions are executed immediately on it so that you can quickly execute your test cases.

 Do not use the mnemonics (the secret seed phrases) generated by these tools on the Ethereum mainnet. Also, do not send ether to the addresses generated with these mnemonics. Doing this, your ether can be lost.

- `truffle develop`: Truffle also has its own built-in blockchain that you can use while running test cases. To start the built-in blockchain instance, run the `truffle develop` command. Once launched, you will see the `truffle` console, and you can run other `truffle` commands from this console without using `truffle` as a prefix before commands. For example, instead of running `truffle compile`, you just run `compile` from the `truffle` console.

Running blockchain nodes

Some official and unofficial blockchain clients are available to download and keep your local blockchain node running on your machine. These blockchain clients need to be synced with the network before running test/deployment scripts:

- **Geth**: Geth (as known as **Go-ethereum**) is an Ethereum blockchain client written in the Go language. This is the official client developed by Ethereum Foundation. You can download it from `https://github.com/ethereum/go-ethereum`.
- **Parity**: Parity Technologies provides the `parity-ethereum` client. This is an unofficial version of the client application that you can download from `https://github.com/paritytech/parity-ethereum`.

- **WebThree**: WebThree is an official client written in the C++ language. You can download it from `https://github.com/ethereum/aleth`.

Some of these blockchain clients also support connecting to Ethereum test networks. Hence, you can connect to a testnet and run your test and deployment scripts there.

Using Infura

Infura is the project that maintains the Ethereum blockchain nodes at their end. In the case of the Geth and Parity clients, you need to install them and keep them running on your machine to sync the blockchain ledger. However, with Infura, you just need to register for free at `https://infura.io` and it provides you with a link and your unique token to use to connect to the blockchain. The benefit of using Infura is that you do not need to run your local blockchain for the mainnet and testnets (Ropsten, Kovan, and Rinkeby), as it is maintained by the Infura service.

To give you an example, MetaMask internally uses an Infura link to connect to the Ethereum blockchain.

You can also use Infura in your `truffle` project. For this, you would need to use the custom provider, `hdwallet-provider`. You can find the usage of this in the `truffle-config.js` file.

Writing contract migration scripts

The migration scripts should be written in the JavaScript file for the `truffle` project. It must be present under the `migrations` folder. As the project progresses in terms of development, your contract also needs changes, and will require contract or setup changes present on the blockchain. This is done by adding new migrations scripts. Using the special contract, `Migrations.sol`, the history of the executed migration scripts are recorded on-chain.

Trigger migration using the migrate option

To initiate migration, you should run the following command:

```
$ truffle migrate
```

This command will run all of the migration scripts, present under the `migrations` folder. The migration scripts are always executed in sorted order.

To maintain this sort order, the developer has to ensure that the script's name always starts with a number. Ensure that it follows the `<Num>_fileName.js` format for a filename. The filename must be prefixed with the number and suffixed with an appropriate name for the file. The following are some examples for filenames:

Script filename	Script number
`1_initial_migration.js`	1
`2_deploymentExample.js`	2
`3_upgrades.js`	3

The migrations script's filename and their script number

The migration starts executing scripts from the first scripts and moves on in increasing order for script execution. The migration script execution is performed in the following manner:

- If all of the migration scripts executed previously and no new migration scripts are added, then migration will not be executed again.
- If the migration command was executed before, it will start the migration from the last file onward and execute the rest of the migration script files.
- However, if the `migrate` command uses `--reset` flag, for example, `truffle migrate --reset`, this will enforce `truffle` to start the migration again from the start.

Sample migration script

We have created a migration script, `3_depositContract.js`, which is present in the `migrations` folder of the project. The following are the contents of the file:

```
var DepositContract = artifacts.require("DepositContract");

module.exports = function(deployer) {
  deployer.deploy(DepositContract);
};
```

When you run the `truffle migrate` command, this script will deploy the `DepositContract` contract instance on the local blockchain.

Using artifacts.require() to get the contract instance

At the start of the migration script, we need to define which contracts instances we would like to interact with. You can get an abstraction instance of a contract as an object using the following example syntax:

```
var contractVariable = artifacts.require("<ContractName>");
```

The following is an example command to get the instance of DepositContract:

```
var DepositContract = artifacts.require("DepositContract");
```

You need to provide a contract name in place of <ContractName>. The contract name you define here should match the contract name defined in the multiple contract files present under the contracts folder. You should not provide a .sol contract filename, because a contract file could contain multiple contracts.

Using module.exports

As shown in the 3_depositContract.js code discussed in the *Sample migration script* section, a migration script must export a function using the module.exports syntax:

```
module.exports = function(deployer) {
   // Deploy contracts using 'deployer'
}
```

This function should have a deployer object as its first parameter in the function. This object provides methods to deploy the contracts on the blockchain network, and saves the status of the deployed contract for later use in the migration and test scripts.
The deployer object should be used for contract deployment for testing, staging, and production. Let's understand how to use the deployer object.

Deployer

Your contract migration files will use the deployer object to stage the deployment tasks for your contracts. You can write your deployment tasks synchronously, as shown in the following example, and they'll be executed in the correct order in which they are specified:

```
deployer.deploy(Contract1);
deployer.deploy(Contract2);
```

First, the deployer will deploy the Contract1 contract, and only then it will deploy the Contract2 contract.

Other than the `deployer` object, there are other optional objects can also be used, such as `network` and `accounts`. Let's look at the `network` object.

Network considerations

The `module.exports` function also initializes the `network` object with the network name that is being used while deploying the contract using migration scripts:

```
module.exports = function(deployer, network) {
  if (network == "development") {
    // Script to be executed for network named "development".
  } else if (network == "rinkeby") {
    // Script to be executed for network named "rinkeby".
  } else {
    // If network name not matched, execute this script
  }
}
```

As you can see in the example, you can check the name of the network and accordingly execute the script or perform a conditional action.

Now, let's look at the third optional object, `accounts`.

Available accounts

You can also get the list of Ethereum accounts provided by the Ethereum client or web3 provider connected to `truffle` while running the migration. `accounts` is an array object, and you can access `accounts` using the index. You can also get the same list of accounts using the `web3.eth.getAccounts()` (this call is present with `web3j.js` version 0.2x.x) function call:

```
var TestContract = artifacts.require("TestContract");

module.exports = function(deployer, network, accounts) {
  deployer.deploy(TestContract, accounts[0]);
}
```

As shown in the preceding example, it deploys a contract named `TestContract` and passes `accounts[0]` (the first Ethereum account) as an argument to the constructor of `TestContract`.

By default, the Ganache GUI and Ganache CLI versions create and initialize only 10 accounts. Hence, using the `accounts` object, you can only get 10 accounts. You can increase the number of accounts using advanced settings in Ganache.

Writing test cases

You can write test cases using different scripting languages, such as JavaScript and TypeScript, as well as in Solidity itself. As a developer, you must ensure that test cases are written to cover at almost 100% coverage. Having 100% code coverage is good enough to ensure that there is a lower probability of bugs in your code. All of the test files should be present in the `test` folder.

To run the test suite, you need to use the following command:

```
$ truffle test
```

You can also specify the single test files to run using the following:

```
$ truffle test ./path/to/test/file.js
```

Now, let's look into how to write the test cases in JavaScript using some available test frameworks.

Writing test cases in JavaScript using Mocha

You can write JavaScript test cases using the Mocha framework and Chai for assertions. You write your test cases for the contract using Mocha, as we can write test cases for other projects, and include it in the `.js` file. These files must be kept in the `test` folder of the project. There is a difference for the Truffle tests when writing tests using Mocha: instead of using the `describe()` function, you should use the `contract()` function. For example, you can see the JavaScript test case file, `DepositContract.test.js`, written for `DepositContract` and present in the `test` folder of the project.

Truffle has a clean-room environment feature you can use:

- Before the execution of the `contract()` function, Truffle will re-deploy all of the contracts again using the migration scripts. This gives each `contract()` function a clean room for the contract states.

- When using different blockchain via Truffle Develop or Go-Ethereum, Truffle starts the migration scripts from the first script. Truffle keeps the migration states of both the network separately.
- The `contract()` function also provides the list of Ethereum accounts in an array object. These accounts are provided by the Ethereum client your Truffle instance is connected with. You can use these Ethereum accounts to initiate the transactions.

The following is the code snippet taken from the JavaScript test case file, `DepositContract.test.js`. As you can see, the `accounts` object is an array of Ethereum accounts:

```
const DepositContract = artifacts.require('DepositContract');
let depositContract;

contract('DepositContract', function (accounts) {
  beforeEach(async function () {
    //For ownable behavior
    this.ownable = await DepositContract.new();
    depositContract = this.ownable;

    //accounts
    owner = accounts[0];
    user1 = accounts[1];
    user2 = accounts[2];
    other = accounts[3];
  });

  describe('as DepositContract', function () {

    it('should fail when deposit 0 ether using depositEther()',
    async function () {
      await expectThrow(depositContract.depositEther({from: user1}),
       EVMRevert);
      const balance = await web3.eth.getBalance(depositContract.address);
      assert.equal(balance, 0);
    });

    it('should deposit 1 ether using depositEther()', async function () {
      await depositContract.depositEther(
        {from: user1, value: web3.utils.toWei("1" , "ether")});
      const balance = await web3.eth.getBalance(depositContract.address);
      assert.equal(balance, web3.utils.toWei("1" , "ether"));
    });
  }
```

As you can see in the preceding code, at the first line, we have initialized the abstract contract object, DepositContract, and, using that, created an instance of the contract in the contract() function. We have also initialized four different wallet accounts using the accounts array, indexed from 0 to 3. In the describe() function, we have specified two test cases that will be executed when the test suite runs.

Writing test cases in Solidity

You can also write test cases in the Solidity contract. For this, there are standards you need to follow:

- The test contract file should be kept under the test folder.
- The Solidity contract filename must be prefixed with Test, for example, TestDepositContract.sol, which you can find with our project source code.
- The name of the test functions present in the contract must be prefixed with test. For example, we have the testDeposit() function defined in TestDepositContract.sol.
- For assertion functions, you need to import the Assert.sol contract file in your test contract. The syntax to import it is import "truffle/Assert.sol".
- To get the address of the deployed contract, you can import DeployedAddresses.sol, as shown in the following code example.
- There are many hooks provided—beforeAll(), beforeEach(), afterAll() and afterEach(). These are the same hooks provided by the Mocha framework, as well for JavaScript tests.

The following is a code snippet from the test case file, TestDepositContract.sol:

```
import "truffle/Assert.sol";
import "truffle/DeployedAddresses.sol";
import "../../contracts/chapter05/DepositContract.sol";
contract TestDepositContract {
    // Truffle sends 3 ether to this contract for testing
    uint public initialBalance = 3 ether;

    DepositContract depositContract;

    function beforeEach() public {
        depositContract =
        DepositContract(DeployedAddresses.DepositContract());
    }
```

```
function testUnpaused() public {
    Assert.equal(depositContract.paused(), false, "Contract should be
    UnPaused");
}
}
```

In the preceding test contract file, we have a `uint public` state variable, `initialBalance`, which is initialized with `3 ether`. When the contract is deployed by Truffle, it will automatically send 3 ether to the contract. As you can see, we are using the `DeployedAddresses` utility contract to get the address of the already-deployed (by migration scripts) `DepositContract`. The `beforeEach()` function will be called before each test function in the contract. The `testUnpaused()` function is a valid test case and Truffle will display the result of the test case in the console when executed.

Debug transactions

Using Truffle framework, you can debug any transaction with Truffle's `debug` command. To run `truffle` in debug mode, there are some requirements that you need to fulfill:

- If you are running Ganache, it should be running in verbose log mode. However, verbose logs are not much helpful in debugging the transactions manually. Its useful for truffle debug mode to read and parse the logs. You can run `ganache-cli` with `-v` flag to run Ganache in verbose log mode.
- The full contract code should be available.
- The ID of the transaction that you want to debug should be known. To find out the ID of a transaction that is not working as per your expectations with test cases, you can stop the test case when it fails and can find transaction information in Ganache. Ganache GUI is sometimes more helpful in finding and debugging the current state of the blockchain.

You can debug a transaction in `truffle` using the following command:

```
$ truffle debug <Transaction_ID>
```

The following is the command to debug a specific transaction:

```
$ truffle debug
0x13b60d1b9cb73ecbbac272fcdc8ecb3d66b50ddf6a2bfcfd40e1765f31155ba6
```

After running the preceding command, `truffle` will compile the code, fetch the transaction data from the Ganache logs, and display the following output:

```
Gathering transaction data...

Addresses affected:
  0xeba7ddf17f97c04377cf3d7f014d9656b243c25d - DepositContract

Commands:
(enter) last command entered (step next)
(o) step over, (i) step into, (u) step out, (n) step next, (;) step instruction
(p) print instruction, (h) print this help, (q) quit, (r) reset
(b) add breakpoint, (B) remove breakpoint, (c) continue until breakpoint
(+) add watch expression (`+:<expr>`), (-) remove watch expression (-:<expr>)
(?) list existing watch expressions
(v) print variables and values, (:) evaluate expression - see `v`


DepositContract.sol:

  9:  * @title Deposit Contract to hold Ether
 10:  */
 11: contract DepositContract is Ownable, Pausable {
     ^^^^^^^^^^^^^^^^^^^^^^^^^^^^^^^^^^^^^^^^^^^^^^^^

debug(development:0x13b60d1b...)>
```

Using Truffle to debug a transaction

As shown in the preceding screenshot, it displays the contract that was affected by this transaction, along with the contract address. It also lists the number of commands that are available in `truffle` debug mode to give you the tools to dig deeper. To debug the next statement, you just press *Enter* on the console and it listed the actual function call that was initiated using this transaction ID:

```
DepositContract.sol:

22:         */
23:         function() external payable {
24:             depositEther();
                ^^^^^^^^^^^^^^^

debug(development:0x13b60d1b...)> █
```

Truffle debug mode: The transaction called the contract's depositEther() function

You can continue debugging the transaction using the debug commands present in debug mode. This tool will be helpful when you are unsure about what caused unexpected behavior in a contract and you will be able to dig deeper with it.

Summary

In this chapter, we looked into various tools and frameworks such as Ganache, Truffle framework, and Infura links. We learned how to use the GUI version of the Ganache application to get your personal Ethereum blockchain up and running quickly for testing and deployment. We also covered how you should choose which Ethereum blockchain client to use according to your differing requirements. We looked into Truffle and how to write test cases and migration scripts.

As of the time of writing, Truffle is the most used development suite by Solidity contract developers, as it supports most of their requirements, although there are some other tools that would provide more detailed information about the Solidity contracts and the overall project. We will look into those tools in the next chapter, where we will cover code quality tools.

In the next chapter, we are going to learn about tools like Surya, Solhint, Solium and use these tools which are mostly used while developing Solidity contracts.

Questions

1. What is the difference between Ganache and the Ganache CLI version?
2. Are there any other applications for a personal Ethereum blockchain?
3. Can we use Truffle framework for production?
4. Can Truffle be used for private blockchains?
5. Is it always better to use Infura?

6
Taking Advantage of Code Quality Tools

Code quality is one of the important aspects of writing applications. Good quality code always tends to have fewer bugs and problems when deployed in production. To maintain and improve code quality, there are always some tools specific to the language you are coding with. Similarly, there are some tools for Solidity as well.

In this chapter, we are going to learn about some of the tools that are used while developing Solidity contracts. There are many open source tools available for the Solidity language, and these include code quality tools such as contract graph generators, linters, and code coverage tools. Using these contract graph generators, you can view how the contracts are linked together; using the linters, you can fix possible bugs, errors, and stylistic errors; and using the code coverage tools, you can discover which part of the code is well tested and which part is not covered by the test cases.

In this chapter, we will cover the following topics:

- Installing and using `surya` commands to generate reports
- Installing and using the `solhint` linter
- Installing and using the `ethlint` (also known as `solium`) linter
- Installing and using the `solidity-coverage` tool

Technical requirements

You can find the code used in this chapter on GitHub at https://github.com/
PacktPublishing/Mastering-Blockchain-Programming-with-Solidity/tree/master/
Chapter06.

In this chapter, we are going to use different code quality tools. For each tool, you can install the latest version of these tools. As of writing this book, the following are the latest version of these tools:

- surya version 0.2.8 and onward: GitHub link of the project https://github.com/ConsenSys/surya
- solhint version 2.0.0 and onward: GitHub link of the project https://github.com/protofire/solhint
- solium version 1.2.3 and onward: GitHub link of the project https://github.com/duaraghav8/Ethlint
- solidity-coverage version 0.5.11 and onward: GitHub link of the project https://github.com/sc-forks/solidity-coverage

To install the preceding tools, you must have Node 10.15.1+ installed.

Using the surya tool

The surya tool is an open source command-line utility tool that is used to generate a number of graphical and other reports. The tool works by scanning through the Solidity smart contract files and can generate inheritance and function call graphs. It also generates a function-specific report in a tabular format. Using all of these generated graphs and reports, a developer can understand smart contract architecture and dependencies by doing a manual inspection.

Let's start by installing the surya tool on your machine.

Installing surya

To install the surya utility tool on your machine, run the following command:

```
$ npm install -g surya
```

Once this is installed, you can use the `surya` command in any of your `truffle` projects or even in the folders containing Solidity files.

`surya` provides the following commands:

- `surya describe`: This generates the contract structure in text format
- `surya graph`: This generates the flow graph of the function calls
- `surya inheritance`: This generates the contract inheritance graph
- `surya dependencies`: This generates the C3 linearization of inherited contracts
- `surya parse`: This parses and generates an **Abstract Syntax Tree (AST)** tree in text format
- `surya ftrace`: This generates the function trace of a given function name
- `surya mdreport`: This generates a tabular description for functions of a contract in the markdown format

You can also run the following command to see the syntax and arguments required by each of the previously listed commands:

```
$ surya --help
```

The command will generate following output explaining all of the supported commands along with its syntax and description:

```
✔ surya --help
surya <cmd> [args]

Commands:
  surya describe <files..>                      show file contracts structure.
  surya graph <files..>                         generate graph of contract function
                                                calls.
  surya inheritance <files..>                   generate graph of contract
                                                inheritance tree.
  surya dependencies <target_contract>          output a linearized list of smart
  <files..>                                     contract dependencies (linerized
                                                inherited parents).
  surya parse <file>                            output AST generated by the parser
                                                for the specified file in a textual
                                                tree format.
  surya ftrace <function_identifier>            output the selected function call
  <function_visibility_restrictor>              trace in a textual tree format.
  <files..>                                     External calls are marked in
                                                `orange` and internal calls are
                                                `uncolored`.
  surya mdreport <outfile> <infiles..>          output a markdown file

Options:
  -h, --help      Show help                                          [boolean]
  -v, --version   Show version number                                [boolean]
```

The supported surya commands

We will explore the usage of each of these commands in the upcoming sections and learn how to generate different kinds of reports using surya. For the sample code to execute the surya commands and generate different reports, we are using the OpenZeppelin 2.1.1 contracts. You can refer to the contracts present in the OpenZeppelin library; alternatively, some of the contracts are also listed in the Chapter06/SuryaTutorial folder. You can execute all of the surya commands listed in this chapter under the Chapter06/SuryaTutorial folder.

Using surya describe

The surya describe <files..> command is used to generate a report in text format, which describes the functions present in the contract. It lists all of the functions present in the contract supplied as the argument. The command does not list the functions of inherited contracts.

The command takes the Solidity contract file path as an argument. The following screenshot shows the output of the surya describe command executed for the Crowdsale.sol contract:

```
$ surya describe contracts/crowdsale/Crowdsale.sol
```

```
✔    surya describe contracts/crowdsale/Crowdsale.sol
+  Crowdsale (ReentrancyGuard)
   - [Pub] <Constructor> #
   - [Ext] <Fallback> ($)
   - [Pub] token
   - [Pub] wallet
   - [Pub] rate
   - [Pub] weiRaised
   - [Pub] buyTokens ($)
   - [Int] _preValidatePurchase
   - [Int] _postValidatePurchase
   - [Int] _deliverTokens #
   - [Int] _processPurchase #
   - [Int] _updatePurchasingState #
   - [Int] _getTokenAmount
   - [Int] _forwardFunds #


($) = payable function
# = non-constant function
```

The surya describe command result for the Crowdsale.sol contract

Let-us understand the symbols and meaning of this output:

- The generated output lists the visibility of the function as follows:
  - **[Pub]**: This is highlighted in green and the function is a `public` function
  - **[Ext]**: This is highlighted in blue and the function is an `external` function
  - **[Int]**: This is highlighted in gray and the function is an `internal` function
- Some symbols are also used:
  - **# (hash)**: The symbol is highlighted in red, indicating that the function can change the contract state variable
  - **($) (dollar sign)**: The symbol is highlighted in yellow color, that the function is `payable` functions and can receive ether as well along with the transaction

Having the preceding report generated we can quickly identify that functions that are changing the contract's state variables and `payable` functions must be checked rigorously to ensure that there are no bugs in them.

Generating an inheritance graph

The `surya inheritance <files..>` command is used to generate the contracts inheritance tree in a graphical format. This command takes the list of files to parse and generates a connected graph with contract names and from which other contracts it's inheriting from.

The command takes the Solidity file's path as an argument and generates the output in a graph description language-compatible (DOT) format. As you can see in the following command, we converted the DOT format language to generate the `.png` image using the `dot` command:

```
$ surya inheritance contracts/token/ERC20/*.sol | dot -Tpng > erc20-
inheritance.png
```

The preceding command will generate a file named `erc20-inheritance.png` in the folder; the generated graph is as follows:

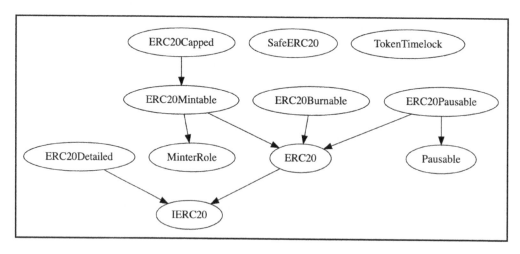

The surya inheritance graph of the contracts present in the contracts/token/ERC20 folder

In the preceding diagram, you will discover the following:

- The libraries in Solidity do not have any connection, as shown in diagram. The library contract cannot have any inheritance. For example, in the preceding diagram, `SafeERC20` is a library contract. No arrow (→) is connected to it.
- The contracts that do not have any connection are not inheriting from any other contract. Alternatively, the files of those contracts are not provided to the `surya` command. For example, in the preceding diagram, the `Pausable` contract also inherits from the `PauserRole` contract; however, this is not shown in the graph generated as the `PauserRole` contract file is not shared to the `surya` command. Hence, `surya` may not pick up all of the files. It explicitly requires all of the files reference in the command itself.
- The contracts that inherit from other contracts are linked together with an arrow (→). The contract that is connected with the end part of the arrow is the inheriting contract, while the arrow connected with the pointed end is the inherited contract. For example, `ERC20Capped` is the inheriting contract that inherits from the `ERC20Mintable` contract, which further inherits from the `MinterRole` and `ERC20` contracts.

Similarly, you can execute the following command to generate the inheritance graph of the contracts present in the ERC721 folder:

```
$ surya inheritance contracts/token/ERC721/*.sol | dot -Tpng > erc721-
inheritance.png
```

This will generate the `erc721-inheritance.png` image file in the folder and the following graph is generated:

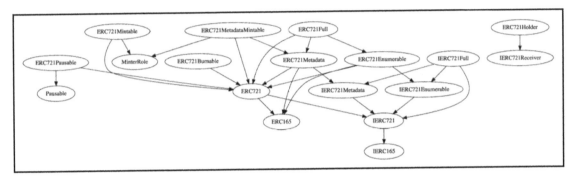

The surya inheritance graph of the contracts present in the contracts/token/ERC721 folder

Generating a function call graph

The `surya graph <files..>` command is used to generate the function call flow graph of the contract. The generated graph also shows the call visibility as an internal call or an external call. The graph only covers the contracts known to the `surya` command. The command takes the contract(s) files as a command argument:

```
$ surya graph contracts/token/ERC20/*.sol | dot -Tpng > all.png
```

The preceding command will generate the function's call flow of all of the contracts present in the `contracts/token/ERC20` folder. The following screenshot is a part of the bigger graph, which we cannot display on a single page.

You can get the full graph generated by this command on GitHub at `https://github.com/PacktPublishing/Mastering-Blockchain-Programming-with-Solidity/blob/master/Chapter06/SuryaTutorial/generated_files/all.png`:

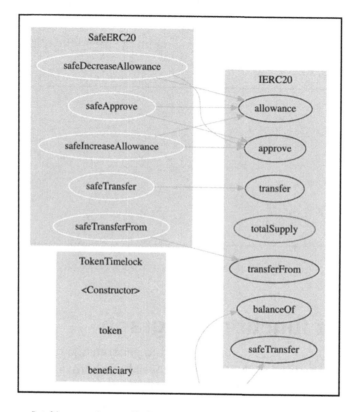

Part of the surya graph generated for the contracts present in the contracts/token/ERC20 folder

Parsing contracts

The `surya parse <file>` command parses the contract file and generates the AST in text format. The contract's parse output is helpful for auditors to understand the flow of the contract. You can parse any contract file as follows:

```
$ surya parse contracts/token/ERC20/IERC20.sol
```

The command only takes the contract file name as an argument. The preceding command will generate the AST parse output of the given `IERC20.sol` contract, as shown in the following screenshot:

```
✔    surya parse contracts/token/ERC20/IERC20.sol
├─ type: SourceUnit
└─ children
   ├─ 0
   │  ├─ type: PragmaDirective
   │  ├─ name: solidity
   │  └─ value: ^0.5.0
   └─ 1
      ├─ type: ContractDefinition
      ├─ name: IERC20
      ├─ baseContracts
      ├─ subNodes
      │  ├─ 0
      │  │  ├─ type: FunctionDefinition
      │  │  ├─ name: transfer
      │  │  ├─ parameters
      │  │  │  ├─ type: ParameterList
      │  │  │  └─ parameters
      │  │  │     ├─ 0
      │  │  │     │  ├─ type: Parameter
      │  │  │     │  ├─ typeName
      │  │  │     │  │  ├─ type: ElementaryTypeName
      │  │  │     │  │  └─ name: address
      │  │  │     │  ├─ name: to
      │  │  │     │  ├─ storageLocation
      │  │  │     │  ├─ isStateVar: false
      │  │  │     │  └─ isIndexed: false
```

A portion of the result generated using the surya parse command for the IERC20.sol contract

The full output of the preceding command can be found on GitHub at `https://github.com/PacktPublishing/Mastering-Blockchain-Programming-with-Solidity/blob/master/Chapter06/SuryaTutorial/generated_files/ERC20_parse.txt`.

As you can see in the preceding screenshot of parse output, it starts with `PragmaDirective` showing which Solidity version is used in the contract file. After that, the `ContractDefinition` starts, where it lists the contract name, `IERC20`. Further down, it keeps listing each function under `FunctionDefinition`, along with its parameters and return type.

Generating function traces

The `surya ftrace` command is used to generate the function call traces. It lists all of the function calls made from the given function. The command takes the contract and the function name as an argument to find its call traces:

```
$ surya ftrace TokenTimelock::release all contracts/token/ERC20/*.sol
```

The preceding command will generate the following output in text format:

```
✔   surya ftrace TokenTimelock::release all contracts/token/ERC20/*.sol
└─ TokenTimelock::release
   ├─ IERC20::balanceOf | [Ext] !
   └─ IERC20::safeTransfer | [Int] 🔒 ●
```

The surya ftrace output for the release() function present in the TokenTimelock contract

As shown in the preceding screenshot, the `release()` function, which is present in the `TokenTimeLock` contract, makes calls to the `IERC20.balanceOf()` and `IERC20.safeTransfer()` functions.

Listing inheritance dependencies

The `surya dependencies <target_contract> <files..>` command lists the C3 linearization of the inheritance for the given contract. The output lists the most derived contract; this means that, if the same function is present in the inherited contracts, it will look for the function in the listed order and use the definition (if present).

In this example of the command, we are using the `0x` project `Exchange.sol` contract. The `0x` project is an open source **Decentralized Exchange (DEX)**. Follow the following commands:

```
$ git clone https://github.com/0xProject/0x-monorepo.git
$ cd 0x-monorepo/contracts/exchange/contracts
$ surya dependencies Exchange **/*.sol
```

The preceding command will generate the following output:

Surya dependencies for the contract's inheritance order

The `Exchange.sol` contract file has a big list of inheritance. You can refer to the `Exchange.sol` code on GitHub at `https://github.com/0xProject/0x-monorepo/blob/development/contracts/exchange/contracts/src/Exchange.sol`.

Following is the code snippet of the contract and its inheritance:

```
contract Exchange is
    MixinExchangeCore,
    MixinMatchOrders,
    MixinSignatureValidator,
    MixinTransactions,
    MixinAssetProxyDispatcher,
    MixinWrapperFunctions
{
    //Rest of the code
}
```

Solidity uses C3 Linearization, which is an algorithm used to find the order in which functions should be inherited in the contract when multiple inheritance is used. The important thing to remember is that the Solidity compiler composes the inheritance graph from right to left. As you can see in the Exchange contract, the rightmost contract it has inherited is MixinWrapperFunctions, and this contract will be looked to first to find a given function. You can also see that this contract is listed first in the output of the surya dependencies executed on the Exchange contract. Hence, the order generated by the surya command is used to find a function in the contracts in the given order starting from the first contract.

Generating the markdown report

The surya mdreport (markdown report) command generates the same report as the surya describe command. The only difference is that the surya mdreport command generates the report in markdown format. The command used to generate the report takes the new report filename and the contract filename, as follows:

```
$ surya mdreport erc20.md contracts/token/ERC20/ERC20.sol
```

The preceding command will generate the report in markdown format; when viewed in the markdown viewer or on GitHub, it looks similar to the following screenshot:

Sūrya's Description Report

Files Description Table

File Name	SHA-1 Hash
contracts/token/ERC20/ERC20.sol	d51fee97e186bd4f036fe4951dc9ee2ecef42ba8

Contracts Description Table

Contract	Type	Bases		
L	**Function Name**	**Visibility**	**Mutability**	**Modifiers**
ERC20	Implementation	IERC20		
L	totalSupply	Public !		NO !
L	balanceOf	Public !		NO !
L	allowance	Public !		NO !
L	transfer	Public !	●	NO !
L	approve	Public !	●	NO !
L	transferFrom	Public !	●	NO !
L	increaseAllowance	Public !	●	NO !
L	decreaseAllowance	Public !	●	NO !
L	_transfer	Internal 🔒	●	
L	_mint	Internal 🔒	●	

The surya markdown report of ERC20.sol viewed in the markdown viewer

You can also find the mdreport file on GitHub location at `https://github.com/ PacktPublishing/Mastering-Solidity/blob/master/Chapter06/SuryaTutorial/ generated_files/erc20.md`. As you can see in the preceding screenshot, the report generates a table with the following columns:

- **Contract**: This is the name of the contract.
- **Function Name**: This is the function name present in the contract.
- **Visibility**: This is the visibility of the function listed; **Public** and **External** visibility are marked with a red exclamation mark to let the developer know to be careful about these methods.

- **Mutability**: This shows whether the function can change the state variables or not. If a function modifies the state variable, it's called a **mutable function**, and these functions are marked by a red octagon in the table.
- **Modifiers**: This shows the list of the custom function modifiers apply on a function. If there is no function modifier present for the public and external function, it will show **NO!** with a red exclamation to make the developer aware of the impact this method may cause.

Understanding Solidity linters

Linters are the utility tools that analyze the given source code and report programming errors, bugs, and stylistic errors. For the Solidity language, there are some linter tools available that a developer can use to improve the quality of their Solidity contracts. These tools report the known pattern of errors or bugs and also check any security flaws that could be checked by the developers to ensure the safety of the contract.

However, these linter tools should be used along with the compiler's reported warnings. Because the compiler itself reports many warnings and informs the developer about the best language guidelines, it also suggests using the improved language syntax to reduce the security bug. You should bear in mind that the compiler warnings are not sufficient to have good quality code. You can also automate and write a script to compile the code and run linters on it after successful compilation.

For the Solidity language, there are two commonly used linter tools available; these are `solhint` and `solium` (also known as `ethlint`). We will examine each of these tools in the next section.

Files related to this section are present in the `Chapter06/LintersTutorial` folder.

Using the solhint linter

The `solhint` linter is an open source Solidity linter created by ProtoFire. `solhint` provides security and style guideline-specific validations. It also validates some of the best practices for Solidity; for example, the code should not have any empty blocks like the following one:

```
function getStatus() public view returns (bool) {
    //There is no code present in this function.
    //Hence, solhint will report issues for this
    //empty function definition.
}
```

The empty block example is just one of the pattern `solhint` linter checks. There are many other patterns that linters check in code.

Installing the solhint linter

You can install the `solhint` linter globally on your machine using the following npm command. After the installation, you can execute the `solhint` command in any folder on your machine:

```
$ npm install -g solhint
```

Once it is installed, you can verify that the installation was successfully done by querying the version of the `solhint` command installed:

```
$ solhint -V
```

The preceding command will return the currently installed version of `solhint`. Once you see the version of `solhint` displayed on your shell or command window, you are good to go and you can use the `solhint` command for your `truffle` or Solidity contract files.

Using solhint

To list all of the available options of the `solhint` command, you can run the following command with the help option:

```
$ solhint -h
```

The command will list all of the options available, as shown in the following screenshot:

```
✔  solhint -h
Usage: solhint [options] <file> [...other_files]

Linter for Solidity programming language

Options:

  -V, --version                           output the version number
  -f, --formatter [name]                  report formatter name (stylish, table, tap, unix)
  -w, --max-warnings [maxWarningsNumber]  number of allowed warnings
  -c, --config [file_name]                file to use as your .solhint.json
  -q, --quiet                             report errors only - default: false
  --ignore-path [file_name]               file to use as your .solhintignore
  -h, --help                              output usage information

Commands:

  stdin [options]                         linting of source code data provided to STDIN
  init-config                             create in current directory configuration file for solhint
```

The solhint command options available

As you can see in the preceding screenshot, you can use the different options of `solhint` to generate the report in a different format, provide the `solhint` configuration file to use, and more.

The only requirement to run `solhint` is its configuration file. The config file named `.solhint.json` has to be present at the location from where you are executing the `solhint` command. Otherwise, the config file address should be provided with the `--config` option. If the configuration file is not present, you can create the file using the `init-config` option, as follows:

```
$ solhint init-config
```

Using the preceding command, a default config file will be created in the same folder.

The following is the sample command that you can run under your `truffle` project directory to lint all of the Solidity files present in the `contracts` folder:

```
$ solhint "contracts/**/*.sol"
```

You can also run the `solhint` linter on a single file by providing the relative path to the contract file that you want to lint:

```
$ solhint contracts/ExampleContract.sol
```

Using both of the preceding commands, the `solhint` linter will run on the files provided and return the result in the console. The result contains a filename, line number, and the rule name that is violated in the Solidity code. The rules that are violated are not compiler errors; these are just complaints that can be fixed by the developer. However, it's the developer's decision to fix or not fix any issue reported by linters. For example, the returned result looks like the following:

```
contracts/ExampleContract.sol
   1:16   error     Compiler version must be fixed                              compiler-fixed
   7:0    error     Definition must be surrounded with two blank line indent    two-lines-top-level-separator
   8:4    warning   Explicitly mark visibility of state                         state-visibility
  12:4    warning   Fallback function must be simple                            no-complex-fallback
  17:16   warning   Variable "hash" is unused                                   no-unused-vars
```

Solhint executed on ExampleContract.sol

In the preceding output, you can see that some errors and warnings are reported by `solhint` for the `ExampleContract.sol` file. The `solhint` report has the following columns:

- The first column shows on which line number the problem is
- The second column shows the intensity of the issue, is this an error or a warning

- The third column shows the issue description
- The fourth column shows the rule name applied

This contract file is intentionally created to showcase linter errors and warnings.

`solhint` has a configuration file named `.solhint.json`. You can configure this file to either include or ignore the rules checked during the linter process.

 Recently, there has been some issues with the `solhint` latest version and Solidity compiler above 0.5.0 onward. If it is not working, use the old version 1.0.15 to lint your code. To install this version use the command `npm install -g solhint@1.0.15`.

Using the ethlint linter

The `ethlint` linter was previously known as `solium`. However, even as of now, the command name is still `solium`. The `ethlint` linter is another linter that is similar to `solhint`. It also analyzes the Solidity code provided and reports the style guideline and security issues. The `ethlint` linter also supports an option to fix the reported issues.

Let's start by installing the `ethlint` linter tool on your machine.

Installing ethlint

To install the `ethlint` linter globally on your machine, run the following command:

```
$ npm install -g ethlint
```

Once it is installed, you can verify that the installation has successfully completed by querying the version of the `ethlint` command installed:

```
$ solium -V
```

The preceding command will return the current version of the `solium` installed. Once you see the version of `solium` displayed on your shell or command window, you are good to go and use the `solium` lint for your `truffle` or Solidity contract files.

Using solium

To list all of the available options with the `solium` command, you can run following command with the help option:

```
$ solium -h
```

The command will list all of the options available, as shown in the following screenshot:

```
✔  solium -h
Usage: solium [options] <keyword>

Linter to find & fix style and security issues in Solidity smart contracts.

Options:
  -V, --version                     output the version number
  -i, --init                        Create default rule configuration files
  -f, --file [filepath::String]     Solidity file to lint
  -d, --dir [dirpath::String]       Directory containing Solidity files to lint
  -R, --reporter [name::String]     Format to report lint issues in (pretty | gcc) (default: "pretty")
  -c, --config [filepath::String]   Path to the .soliumrc configuration file
  -, --stdin                        Read input file from stdin
  --fix                             Fix Lint issues where possible
  --fix-dry-run                     Output fix diff without applying it
  --debug                           Display debug information
  --watch                           Watch for file changes
  --hot                             (Deprecated) Same as --watch
  --no-soliumignore                 Do not look for .soliumignore file
  --no-soliumrc                     Do not look for soliumrc configuration file
  --rule [rule]                     Rule to execute. This overrides the specified rule's configuration in soliumrc if present (default: [])
  --plugin [plugin]                 Plugin to execute. This overrides the specified plugin's configuration in soliumrc if present (default: [])
  -h, --help                        output usage information
```

The available solium command options

As you can see in the preceding screenshot, all of the options are supported by the `solium` command. Let's gain an understanding of a few of the important ones, as follows:

- `-i, --init`: This creates the default `solium` configuration file in the current folder. Later, you can configure it according to the requirements of the different rule for your project.

- `-f, --file`: This option takes a single Solidity file as an argument and runs the linter on the file.

- `-c, --config`: This option allows you to pass the configuration file path. This option should be used if the configuration file is not present in the current folder.

- `--fix`: The `solium` tries to fix some of the issues automatically in files if this option is used. For non-fixed issues, you will have to fix them manually.

- `--watch`: With this option, you can continue watching a file or files for any changes. If the file is changed, the linter will trigger automatically.

The requirement of the `solium` command is the configuration file. The config files named `.soliumignore` and `.soliumrc.json` have to be present in the location where you are executing the `solium` command from. Otherwise, the config file address should be provided with the `--config` option. If the config file is not present, you can create the file using the `--init` option, as follows:

```
$ solium --init
```

The preceding `init` command will create the `solium` configuration files, `.soliumignore` and `.soliumrc.json`, in the current folder with the default configuration settings:

- `.soliumignore`: This configuration file contains the names of the files and directories to ignore during the linting process.
- `.soliumrc.json`: This file contains the configuration that instructs `solium` how to lint your project. You can specify the linting rules that you will require in your project.

To run the `solium` linter on all of the Solidity files present in the folder, you can run the following command specifying the folder name:

```
$ solium -d contracts
```

We run the preceding command under the `Chapter06/LintersTutorial` folder. The project contains an example Solidity contract called `ExampleContract.sol`. This contract is only added to show the linter's output:

```
✔    solium -d contracts

contracts/ExampleContract.sol
   14:8     warning     Provide an error message for require().              error-reason
   17:8     error       Variable 'hash' is declared but never used.          no-unused-vars
   30:8     warning     Provide an error message for require().              error-reason
   33:24    warning     Consider using 'transfer' in place of 'send'.        security/no-send
   34:8     warning     Provide an error message for require().              error-reason
   38:4     error       "test": Avoid assigning to function parameters.      security/no-assign-params

✖ 2 errors, 4 warnings found.
```

The solium command executed in the contracts folder

As you can see in the preceding screenshot, the warnings are shown in yellow highlighted text and errors are shown in red highlighted text. It also shows the contract filename in which the issue is found, along with the line number. In the last column, it shows the rule name it violates.

You can also use the single file command option of `solium`. You can run the command as follows:

```
$ solium -f contracts/ExampleContract.sol
```

The preceding command will also generate the same output as the previous screenshot.

The solidity-coverage tool

Code coverage tools are used to determine which part of the code is covered and tested by the different test cases and which part is not tested. These tools offer an insightful view of the code and its related test cases. Once developers write the test cases for their Solidity project, they can use the coverage tools to find their code coverage. The more the code is covered with test cases, the lower the probability that you will find any bugs in the code in the future.

For Solidity, there is an open source tool called `solidity-coverage`. In this section, we will take a look at this tool in more detail and use it on the sample code. Files related to this section are present in the `Chapter06/LintersTutorial` folder.

Installing solidity-coverage

You can install the `solidity-coverage` tool in your `truffle` project by running the following command:

```
$ npm install --save-dev solidity-coverage
```

This command will install the latest version of the `solidity-coverage` tool in the project only; the command will not be accessible globally. Once it is installed, you should be able to run the command that we have listed in the next section, as there is no way to check whether the installation of the command was successful.

Using solidity-coverage

The requirement for using `solidity-coverage` tools is that your project must have the test cases present in the `truffle` project. This is because all of these test cases will be executed and the corresponding Solidity code will be instrumented by the tool.

The following is the command to run the `solidity-coverage` tool in the `truffle` project. If this reports any error, ensure that your test cases are working with the `truffle test` command, because this command will be initiated by the `solidity-coverage` tool internally:

```
$ ./node_modules/.bin/solidity-coverage
```

The preceding command will internally start the `testrpc` instance (`testrpc` is the old name of `ganache-cli`) on port number `8555`, run all of the test cases present for the `truffle` project, and instrument the tests for the Solidity code. Our `truffle` project is configured to use the `8555` port in the `truffle-config.js` file so that `solidity-coverage` can run directly without any configuration needs. However, if your project is using any other port, it is required that you run Ganache on that port. Once the command execution completed, it will generate the coverage report in the console itself:

File	% Stmts	% Branch	% Funcs	% Lines	Uncovered Lines
contracts/	39.13	40	37.5	41.67	
DepositContract.sol	90	100	75	90.91	53
ExampleContract.sol	0	0	0	0	... 33,34,39,40
All files	39.13	40	37.5	41.67	

The Solidity coverage report in the console

In the preceding screenshot, you can see that, in the `contracts` folder, there are two files. For the `DepositContract.sol` file, the test coverage covers 90% of statements, 100% of branches, 75% of functions, and 90% of the lines. However, for the `ExampleContract.sol` file, there is 0% coverage.

Now, let's find out exactly which part of the code is not tested. For this, the `solidity-coverage` tool also generates a folder named `coverage`, which will contain the HTML-formatted report. In this folder, open the `index.html` file; it will give you the formatted coverage result, as shown in the following screenshot:

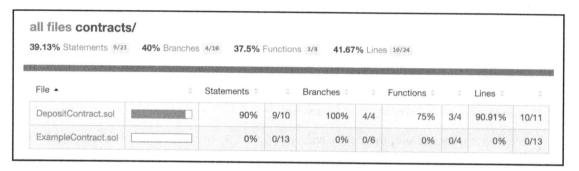

The Solidity coverage report in HTML format

As you can see in the preceding screenshot, `DepositContract.sol` is shown in green. This means that the contract is well-covered. However, the `ExampleContract.sol` file does not have sufficient code coverage.

Once you open each of the contract files, you will be able to see which part of the code is covered by the test cases and which part of the code is not covered. The following screenshot shows a sample snippet of the `DepositContract.sol` contract:

```
41          function withdraw(uint _amount) public {
42    7x          require(balances[msg.sender] >= _amount);
43    5x          balances[msg.sender] = balances[msg.sender].sub(_amount);
44    5x          totalDeposited = totalDeposited.sub(_amount);
45    5x          msg.sender.transfer(_amount);
46    5x          emit Withdrawn(msg.sender, _amount);
47          }
48
49          /**
50           * @dev Withdraw all the deposited ether balance of the user.
51           */
52          function withdrawAll() public {
53              withdraw(balances[msg.sender]);
54          }
```

The code coverage highlighted in the code

As shown in the preceding screenshot, the `withdraw()` function is completely covered by the test cases, as it's showing the lines with green numbers on the left. These numbers signify how many times that particular line has been called. Additionally, the `withdrawAll()` function is not covered by the test cases and is highlighted in red.

It is a good practice to write enough test cases to cover at most 100% code coverage. If you have 100% code coverage, your contracts would have less probability of having bugs. However, this is not guaranteed that 100% coverage ensures that there is no bug present in the code.

Summary

In this chapter, we explored the `surya` tool, which is helpful in generating many types of reports such as inheritance graphs, function flow graphs, inheritance dependencies, and function traces. Following this, we learned about the installation and usage of Solidity linters, such as `solium` and `solhint`, and how you can improve the code quality and find security issues. We learned about the installation and usage of code coverage tool to generate the coverage report.

Using all of these tools is recommended to improve the quality of your code. Once you generate reports with these tools, sometimes, you tend to find the bugs while looking at the reports itself. Hence, it is a good practice to generate these reports and improve the code coverage of your code.

In the next chapter, we are going to look into the most frequently used ERC20 token standard of the Ethereum blockchain. We will deep dive into ERC20 standard functions.

Questions

1. Do linters help to fix security issues?
2. Does 100% Solidity coverage ensure that the testing of the contract is done efficiently?
3. In which cases are function traces useful?
4. Should every Solidity project use linters?

Section 3: Mastering ERC Standards and Libraries

Different ERC standards are present in the Ethereum blockchain. The most famous ones are the ERC20 and ERC721 standards, which we will cover in detail. Some pretested code libraries are also maintained by OpenZeppelin, and we will have a detailed look at understanding those library files.

The following chapters will be covered in this section:

- Chapter 7, *ERC20 Token Standard*
- Chapter 8, *ERC721 Non-Fungible Token Standard*
- Chapter 9, *Deep Dive into the OpenZeppelin Library*
- Chapter 10, *Using Multisig Wallets*
- Chapter 11, *Upgradable Contracts Using ZeppelinOS*
- Chapter 12, *Building Your Own Token*

ERC20 Token Standard

7

Ethereum Request for Comment (ERC) has a standard process to initiate improvement proposals and make them standard. ERC20 is a token standard that's used to create the fungible tokens on Ethereum blockchain—just like **bitcoin** (symbol: **BTC**) and **ether** (symbol: **ETH**) are coins and have economic values. Similarly, by using the ERC20 standard, you can mint as many tokens of a specific type, which also have economic values. This standard allows anyone to easily create their own token on a blockchain.

In this chapter, we are going to look into the ERC20 token standard of the Ethereum blockchain. There are many standards present in Ethereum; however, the most used and popular one is the ERC20 standard. We will understand the API methods of the standard, and we will look into the implementation of these API methods. We will understand when to use ERC20 standards for token creation.

We will cover the following topics in this chapter:

- Overview of the ERC20 standard
- ERC20 standard API functions
- The ERC20 standard code implementation
- Events that are used in the ERC20 standard
- Optional methods of the ERC20 standard
- Advanced methods that can be added to the ERC20 implementation

Technical requirements

You can refer to the code associated with this chapter by going to this book's GitHub repository at `https://github.com/PacktPublishing/Mastering-Blockchain-Programming-with-Solidity/tree/master/Chapter07`.

Overview of the ERC20 token standard

ERC stands for Ethereum Request for Comment. For any technical improvements of smart contract standards, an ERC request is proposed. ERC20 is the technical standard of smart contracts for implementing a *token* on the Ethereum blockchain; 20 is the number of the proposal request. There are multiple steps an ERC goes through to become a standard. An ERC request is first drafted, and then goes through iterations to be updated, accepted, or finalized to become a final **Ethereum Improvement Proposal (EIP)**. You can find all of the standard ERC requests that has been converted into EIPs on GitHub at `https://github.com/ethereum/EIPs/tree/master/EIPS`.

Before moving further, let's understand the difference between a *coin* and a *token*:

- **Coin**: A native digital asset or cryptocurrency of a blockchain is called a **coin**. For example, the bitcoin blockchain has its native cryptocurrency asset, **bitcoin** (symbol: **BTC**). The Ethereum blockchain has its native cryptocurrency asset, **ether** (symbol: **ETH**), which is required in the blockchain to perform any transaction.
- **Token**: A digital asset or cryptocurrency that is built on top of an existing blockchain is called a **token**. For example, **Maker** (symbol: **MKR**) and **Augur** (symbol: **REP**) are ERC20-compliant tokens and are built on the Ethereum blockchain using Solidity smart contracts.

Now, we've understood the difference between the terminology **Coin** and **Token**. However, both of these have some economic values associated with them. For example, one ether coin was worth more than $1,200 on 11th January 2018. Similarly, one Maker token was worth more than $1,000 on same day. These are also called **crypto assets**, and they are being traded on cryptocurrency exchanges. These cryptocurrency exchanges are just like stock exchange (to trade shares), but on these exchanges only these cryptocurrencies are traded. There are two types of exchanges:

- **Centralized Exchange**: On a centralized exchange, you have to send your cryptocurrency coin or token to the exchange's account. Then, they allow you to trade on their platform. As the coins and tokens are held on an exchange's account, there might be a trust issue. It is also possible that if the exchange's account is hacked, you will lose your cryptocurrencies. Binance and Bitfinex are some examples of centralized exchanges. Also there have been some attacks that happened in the past on some centralized exchanges; one of the infamous attack was on Mt. Gox which happened in the year 2014 in which $400+ million worth of bitcoins were stolen.

- **Decentralized Exchange**: On a decentralized exchange, you do not have to send your cryptocurrency coin or token to exchange's account. Your coins are kept in your wallet only. Instead, you directly trade via exchange platform; coins aren't even kept on the exchange platform. They are instantly transferred from one person's wallet to another person's wallet. Using decentralized exchanges, you are safe from hackers as you are not keeping your cryptocurrency on the exchange's account. KyberNetwork, IDEX, and EtherDelta are some of the decentralized exchanges that are available for P2P trading.

 Ethereum has its own native currency, ether, which is not ERC20-compliant. However, on the Ethereum blockchain, to perform transactions for any ERC20 token, you would need to have ether balance to pay for the transaction fee.

Use of ERC20 in crowdfunding

Crowdfunding is also known as **crowdsale**. In recent years, using blockchain technology, many projects have done crowdfunding and received funding in cryptocurrencies such as bitcoin and ether. In return, these projects give tokens to investors. Ethereum itself got funded using crowdfunding in bitcoin back in the year 2014, and its investors received ether in 2015 when Ethereum released its first version.

However, before the Ethereum blockchain, crowdfunding was done manually and tokens were distributed later. Also, it was difficult to do crowdfunding using cryptocurrencies. With the Ethereum blockchain and especially with the ERC20 standard, using the power of smart contracts, it becomes very easy for any project to receive the funding in ether and instantly generate the tokens for the investors. The investor is also able to trade those ERC20 tokens on the cryptocurrency exchanges as the tokens have some value.

This way of doing crowdfunding/crowdsale became popular with the name **Initial Coin Offering (ICO)**, **Security Token Offering (STO)** and **Initial Exchange Offering (IEO)**. Companies or projects looking for funding announce their ICO and receive the funding in ether on a special crowdfunding smart contract and mint the new ERC20 standard tokens for the investor. This process is done in a decentralized manner; hence, it does not require manual intervention and is performed without any middle man.

In this chapter, we will discuss the ERC20 token standard. We will look at crowdfunding-related smart contracts in `Chapter 9`, *Deep Dive into the OpenZeppelin Library*.

The motivation behind the ERC20 standard

ERC20 defines the standard API for token transfers. When all of the tokens that have been created on the Ethereum blockchain follow the same standard APIs, it becomes easy for different web and mobile cryptocurrency wallets to support these tokens. The cryptocurrency exchanges support trading of tokens on their exchange. If all of the tokens support this ERC20 standard, it would be easy for these exchanges to integrate and it would support trading.

Apart from the cryptocurrency wallets and exchanges, it is also easy for a decentralized exchange to support these standard tokens as they would have to call the ERC20 standard APIs from their smart contracts. For example, EtherDelta, IDEX, and KyberNetwork are some decentralized exchanges built on top of Ethereum blockchain and they support trading of ERC20 tokens.

ERC20 standard API

ERC20 is the token standard API that defines functions so that they can access the status of the token, the token's details, and account balances. You can find the original EIP of the ERC20 standard on the GitHub repository of the Ethereum project: `https://github.com/ethereum/EIPs/blob/master/EIPS/eip-20.md`.

The following is the standard API interface defined in Solidity:

```
interface ERC20FullInterface {
    //Below are the OPTIONAL functions of API
    function name() external view returns (string);

    function symbol() external view returns (string);

    function decimals() external view returns (uint8);

    //Below are the functions an implementation MUST have
    function transfer(address to, uint256 value)
      external returns (bool);

    function approve(address spender, uint256 value)
      external returns (bool);

    function transferFrom(address from, address to, uint256 value)
      external returns (bool);

    function totalSupply() external view returns (uint256);
```

```
function balanceOf(address who) external view returns (uint256);

function allowance(address owner, address spender)
  external view returns (uint256);

event Transfer(
  address indexed from,
  address indexed to,
  uint256 value
);

event Approval(
  address indexed owner,
  address indexed spender,
  uint256 value
);
}
```

In the preceding code, we have shown `ERC20FullInterface`, which also includes optional functions as well. However, in practice, the ERC20 interface does not contain these optional functions. In our GitHub code base, you will find both files: `ERC20FullInterface.sol` (you should not use this interface for your contracts) and `ERC20Interface.sol` (you can use this interface for your contracts).

ERC20 implementation

The ERC20 standard only defines the interface APIs—the implementation should be written according to your needs. There are different ways of using the implementations that are available. The most updated and best place to look for the ERC20 implementation is the OpenZeppelin library of Solidity smart contracts. We have taken a part of the implementation from the OpenZeppelin library and will use it in this chapter.

There are some implementations present without using the SafeMath library. However, we recommend using the version with SafeMath library, as it's easy to understand and integer overflow will not be caused when using the SafeMath library for mathematical operations. Whenever you are writing a contract that has mathematical calculations, ensure that you use the SafeMath library in order to write bug-free contracts.

To give you an overview, the SafeMath library is the Solidity library contract that provides basic functions such as multiplication, division, addition, and subtraction. When the mathematical operation is performed using this library, your contract would be safe against the integer overflow and underflow attacks. You will learn more about integer overflow or underflow attacks in `Chapter 14`, *Tips, Tricks, and Security Best Practices*.

Now, let's understand the contract state variables present in the ERC20 contract implementation.

Contract state variables

In the implementation of ERC20, some basic state variables are needed. These state variables store the balance of each Ethereum account, the approved number of tokens and addresses, and the total supply of tokens.

Let's look at each of the state variables defined in the ERC20 token standard. You can refer to the ERC20 implementation code on GitHub at `https://github.com/PacktPublishing/Mastering-Solidity/blob/master/Chapter07/contracts/ERC20.sol`.

The balances variable stores account balance

The `balances` state variable is defined in the contract as follows:

```
mapping(address => uint256) internal balances;
```

The `balances` variable is defined as `mapping` from `address` to `uint256`, which stores the number of tokens each address has. A mapping contains a key-value pair, where the key is unique, hence there will be only one entry present in the mapping per address. As per the Solidity `mapping` data type, by default, all of the addresses will have 0 (zero) balance. Because the balances of each address are stored in the mapping, its retrieval has `0(1)` time complexity, which is ideal.

The entry in the `balances` mapping is updated when the `transfer()` or `transferFrom()` function is called. To perform token transfers, the ERC20 standard implementation must ensure that the `from` account must have the required token balance to transfer to the `to` account.

In the preceding code, the `balances` state variable has `internal` visibility. Hence, all of the contracts implementing the ERC20 implementation can also read or change the `balances` mapping entry.

The `balanceOf()` function in the ERC20 standard reads the mapping data from the `balances` mapping and returns to the caller.

For example, the `balances` mapping stores the following sample data, as shown in the following table:

Address (owner)	uint256 (balance)
0x956e...795c	1000
0xf3a0...e1e3	200
0x8e5c...7aeb	450
0x9f84...08e6	1350

In the preceding table, where the addresses are truncated, only the first four and last four characters of an address are shown. As shown in the preceding table, each address listed has some token balance which is stored in `balances` mapping.

The allowed variable stores approved balances

The `allowed` state variable is defined in the contract as follows:

```
mapping (address => mapping (address => uint256)) internal allowed;
```

`allowed` is a nested mapping of two Solidity `mapping` data types. In the ERC20 standard specification, it is possible for a token holder, X, to assign some allowances to another account, Y, so that Y is allowed to take the approved number of tokens from the token balance of X:

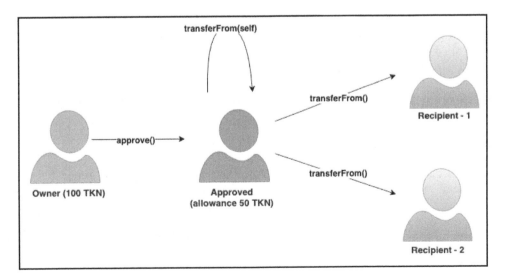

Sample approval flow

As in the preceding diagram, **Owner** has **100 TKN** ERC20 tokens, out of which he has given approval of **50 TKN** to **Approved** person. At any time the **Approved** person can withdraw these **50 TKN** tokens from the **Owner's** account and can further send these tokens to either himself, **Recipient-1** or **Recipient-2** using the `transferFrom()` function. He cannot transfer more than **50 TKN** in total from the owner's account.

For example, the allowances for different addresses is shown in the following table:

Address (from)	Address (spender)	uint256 (allowance)
0x956e...795c	0x1289...a7ad	250
0x956e...795c	0x76c7...ad5a	1000
0x8e5c...7aeb	0xc295...c4c0	100
0x9f84...08e6	0xffe8...9968	400

As shown in the preceding table, the address (from) 0x956e...795c has defined allowances for two different addresses (spender)—0x1289...a7ad and 0x76c7...ad5a. Hence, both of the spender addresses are allowed to take approved allowances (an approved number of tokens) from the 0x956e...795c address.

The totalSupply variable stores the total token supply

The `totalSupply` state variable is defined in the contract as follows:

```
uint256 internal totalSupply_;
```

The `totalSupply` state variable is of `uint256` type. This stores the sum of all of the tokens that exist. This variable gives you an idea of how many of these types of tokens have been generated in total and are held by EOA addresses.

You can also define the `totalSupply` state variable with the visibility of the variable set to `public`, as follows:

```
uint256 public totalSupply;
```

As we know, in Solidity, for the public state variables, the compiler auto-generates getter methods. In the same way, defining `totalSupply` as shown in the preceding code will auto-create a public `totalSupply()` view function, as follows:

```
function totalSupply() public view returns (uint256) {
    return totalSupply;
}
```

However, you would not see the preceding function generated in your contract—it will be generated at compile time.

As shown in the sample mapping data of the `balances` state variable from the preceding table, we can find that the `totalSupply` variable would store 3,000 as the sum of all balances. Let's now look into the ERC20 functions.

The transfer function

The `transfer()` function of the ERC20 standard is used to transfer the tokens from the owner of the token to some other address. As per the standard, the `transfer()` function must emit the `Transfer` event on a successful transferal of tokens. The transaction should revert if `msg.sender` does not have enough tokens to spend. A transfer of a 0 (zero) must also be treated as a valid transfer, and must log the `Transfer` event. The address that's calling the function must have sufficient tokens to transfer to another address.

The ERC20 standard API is as follows:

```
function transfer(address to, uint256 value) external returns (bool
success);
```

The `transfer()` function takes two parameters:

- `to`: This is the Ethereum `address` type to which the tokens must be transferred. We can also refer to this as the destination address.
- `value`: This is the number of tokens to be transferred to the destination address.

The function returns a `bool` value representing whether the transfer of the token is executed successfully or not.

The following is the standard implementation of the `transfer()` function:

```
function transfer(address _to, uint256 _value) public returns (bool) {
    require(_value <= balances[msg.sender]);
    require(_to != address(0));

    balances[msg.sender] = balances[msg.sender].sub(_value);
    balances[_to] = balances[_to].add(_value);
    emit Transfer(msg.sender, _to, _value);
    return true;
}
```

As per the preceding implementation, you can see that it checks that the destination address is not `address(0)`. The `address(0)` is used when performing token minting or burning. If tokens are sent to `address(0)`, they would be burned; therefore, this check is in place to prevent a user from unintentionally burning their token. It also verifies that the `_value` is less than or equal to the balance of `msg.sender`.

If you are calling the `transfer()` function from within a contract, it is recommended to enclose the call with `require()` to ensure that the token transfer executed successfully. In the case of any transfer failure, the transaction should revert. It is defined as follows:

```
require(ERC20.transfer(to, value));
```

The preceding call would require a sufficient token balance to be available in the contract itself. Otherwise, the preceding function call would fail the transaction.

There are a few things that a smart contract developer *must* know about the ERC20 `transfer()` functions. We will go over these in the following section.

Difference between the ether and token transfer functions

The developers must know that there are two types of `transfer()` functions. Developers should not be confused with this and use the required functions when needed.

One is used on the `address` data types to transfer the ether to that address from the contract, as in the following example:

```
uint amount = 1 ether; //Could be any amount to transfer
address(toAddress).transfer(amount);
```

The preceding call would transfer 1 ETH from the contract to `toAddress`. One thing to note is that, before Solidity version 0.5.0, the recipient address should be defined just as an `address` data type. However, since 0.5.0 version onwards, an address receiving ETH must be defined as `address payable`.

The second type is the function defined by the ERC20 standard. This function should be called on the contract with the ERC20 standard API that's supported, as shown in the following example:

```
ERC20(TKNContractAddress).transfer(toAddress, amount);
```

In the preceding call, TKNContractAddress is an address type variable. ERC20() is the contract type. ERC20(TKNContractAddress) casts an address type to a contract type, and then the ERC20 contract APIs would be accessible and can be called. The preceding call would transfer TKN types of tokens from the contract to toAddress.

In case toAddress is a contract, it would receive TKN tokens silently without sending a notification or callback function to any of the contract functions. Let's understand this behavior in detail.

Token transfer does not notify the contact

If you have a contract that should trigger another action once ETH is received on the contract, you can write a fallback function in the contract and call the appropriate function or event to notify it. However, with the ERC20 token standard APIs, it is not possible to get notifications of another action being triggered when an ERC20 standard-specific token is received at the deployed contract. The token transfers are received at the contract silently.

There are some other standards that can be used for this purpose. The ERC223 standard provides the methods that will be called once the tokens are transferred. However, the ERC223 standard is not fully backward compatible with ERC20. Very few companies have used the ERC223 standard since there are some issues with it. Therefore, the developer must know the implications of using ERC223 while architecting their contracts.

Tokens can be locked

As per the ERC20 standard implementation you have seen, the transfer() method does not restrict sending the same tokens to the token contract address itself. This means sending TKN tokens to the TKN ERC20 contract itself. Hence, if the tokens are transferred to the contract address itself, those tokens will be locked forever because the contract code does not have a way to take the tokens out of the contract.

In the past, there have been many ERC20 contracts deployed where the tokens of the same contract are sent to the contract address itself; for example, **Qtum** (symbol: **QTUM**) and **OmiseGo** (symbol: **OMG**) tokens. Mostly, the tokens were sent by human mistake; however, it was the loss of the people who mistakenly sent tokens to the contract address.

In the OpenZeppelin's ERC20 standard implementation, the `transfer()` and `transferFrom()` functions also do not have a check for the contract address itself, such as `require(to != address(this))`. There is a reasoning behind this—to reduce the gas cost of the transaction, as there would be millions of transactions performed using these functions. However, it is the contract developer's responsibility to ensure the safety of their user's funds.

It is not recommended to change the ERC20 implementation and add the contract address check in it. However, the functions related to reclaiming tokens can be used in your ERC20 implementation. This would enable the reclaiming of your contract-specific tokens, as well as other ERC20-specific tokens that were mistakenly sent to the contract.

The transfer transaction details

There are some differences in the transaction details of the ETH transfer transaction and ERC20 token transfer transaction. The developers should know about this and not get confused when looking at the transactions on the block explorer.

The following screenshot, taken from `etherscan.io`, shows the ether transaction where the **From** address sent the ether directly to the **To** address. In this case, the ether balance of the **To** address will be increased:

TxHash:	0x24773589f010dc8c2e136d004f5f56a482bb573c34f2605b17c46f81126e75c2
TxReceipt Status:	Success
Block Height:	7106278 (2 Block Confirmations)
TimeStamp:	46 secs ago (Jan-22-2019 12:29:21 AM +UTC)
From:	0x07f07849760f6303416632558f0ff922cb3158b7
To:	0xfe001e973e1cfce58e4c9ef073ca10a7dbbbaa74
Value:	14.709874 Ether ($1,722.97)

Ether transfer transaction details

For the preceding ether transfer transaction, the transaction data was sent empty.

The following screenshot has been taken from `etherscan.io`, and is showing an ERC20 standard token called **Maker** (symbol: **MKR**) being transferred. As you can see, MKR tokens are sent from the **From** address to the address starting from `0x79d4`. However, the transaction details show that the transaction is performed on the **To** address, which is the address of the MKR token contract. The MKR token contract address is `0x9f8f...79a2`:

TxHash:	0x6eb398f04835018be607db41d10fe3783559252d894c259bad52740018962a91
TxReceipt Status:	Success
Block Height:	7105730 (540 Block Confirmations)
TimeStamp:	2 hrs 30 mins ago (Jan-21-2019 09:56:56 PM +UTC)
From:	0x9c2dfd5acbcf22cb9a82eaaee95824faf5e1e035
To:	Contract 0x9f8f72aa9304c8b593d555f12ef6589cc3a579a2 (MakerToken) ✓
Tokens Transfered:	‣ From 0x9c2dfd5acbcf22c... To 0x79d44f45e2313af... for 3.048 ($1,308.72) ⚄ MKR
Value:	0 Ether ($0.00)

Maker token transfer details

This transaction is a function call to a contract. This is the reason why the **To** address is the contract address. The transaction data that's passed for the preceding transaction is as follows:

```
0xa9059cbb
00000000000000000000000079d44f45e2313afc3d72648aafc1a6a657eaac7e
0000000000000000000000000000000000000000000000002a4cabd9db640000
```

If we decode the preceding transaction data, the parameters are as follows:

- `a9059cbb`: This is the first 4 bytes of the function signature of the `transfer()` function
- `0000...79d44f45e2313afc3d72648aafc1a6a657eaac7e`: The first parameter to the `transfer()` function is the **To** address, `79d44f45e2313afc3d72648aafc1a6a657eaac7e`
- `0000...2a4cabd9db640000`: The second parameter to the `transfer()` function in `uint256` is the amount, `3048000000000000000` (converted from hex to integer).

The approve function

An account (address/contract) can call the `approve` function to approve the transfer of a certain amount to another account (address/contract).

Let's understand this using an example—Alice has 5,000 **TokenX** (symbol: **TKX**) in her Ethereum wallet account. Alice would like to allow her friend, Bob, to spend 1,000 TKX tokens, which Alice has approved for Bob's account. Once the tokens are approved, Bob can take 1,000 TKX or fewer tokens any time from Alice's account and transfer 250 to his friend, Charlie. To perform this operation, Bob will call the `transferFrom()` function, for example, `transferFrom(Alice, Charlie, 250)`, to deduct 250 TKX from Alice and credit these to Charlie's account. After this transaction, Bob is now allowed to take remaining 750 TKX from Alice's account.

The ERC20 standard API is as follows:

```
function approve(address spender, uint256 value) external returns (bool
success);
```

The `approve()` function takes two parameters:

- `spender`: The spender's address to whom the approval will be given and who will be allowed to spend the funds
- `value`: The number of tokens the spender is allowed to fetch from the sender's token balance

The function returns a `bool` value, representing whether the `approve()` function of the token is executed successfully or not. Following is the implementation:

```
function approve(address _spender, uint256 _value) public returns (bool) {
    allowed[msg.sender][_spender] = _value;
    emit Approval(msg.sender, _spender, _value);
    return true;
}
```

As shown in the preceding code it just updates the `allowed` mapping and emits `Approval` event. If you are calling the `approve()` function from within a Solidity contract, it is recommended to enclose the call with `require()` to ensure that the token approval executed successfully. In the case of any failure, the transaction should revert. It is defined as follows:

```
require(ERC20.approve(spender, value));
```

Front-running attack on the approve function

The implementation of the `approve()` function that we looked at in the previous section is prone to front-running attacks. First, understand what is the front-running attack.

An attacker who initiates a transaction which is to be executed before a specific pending transaction that could benefit an attacker financially is called a **front-running attack**.

On the Ethereum blockchain, the transaction gets executed based on the `GasPrice` someone is offering to process their transaction. The more `GasPrice` you provide for a transaction, the more likely it is that your transaction will be executed and added in the block. Also, anyone can read the transactions that are still pending and waiting to be executed in the transaction pool since Ethereum is a public blockchain. Using these properties of the Ethereum blockchain, an attacker can perform a front-running attack. An attacker would observe the transactions that are still pending in the transaction pool and trigger some transaction that could benefit them.

Let's understand how this attack is possible on the `approve()` function:

- **Transaction 1**: Alice calls the `approve(Bob, 1000)` function to allow Bob for 1,000 tokens. This transaction is executed.
- Alice later finds that the 1,000 tokens are too little; he actually wanted to allow 1,500 tokens.
- **Transaction 2**: Alice calls the `approve(Bob, 1500)` function to allow Bob for 1,500 tokens.
- While **Transaction 2** is in a pending state and present in the transaction pool, Bob will see this transaction and initiate another transaction immediately.
- **Transaction 3**: Bob calls the `transferFrom(Alice, Bob, 1000)` function. For this transaction, he sets a high enough `GasPrice` than **Transaction 2** so that this transaction gets executed before **Transaction 2**. This is the transaction that does the front-running attack.
- Now, **Transaction 3** is executed first, and Bob receives 1,000 tokens from Alice's account.
- Now, **Transaction 2** is executed later, which allows Bob to get 1,500 more tokens.
- **Transaction 4**: Bob immediately calls the `transferFrom(Alice, Bob, 1500)` function to get 1,500 more tokens from Alice's account.
- Originally, Alice wanted Bob to get 1,500 tokens only. However, using the front-running attack, Bob ended up getting, in total, 2,500 tokens.

You can also refer `Chapter 14`, *Tips, Tricks, and Security Best Practices*, for detailed understanding of front-running attack.

Preventing a front-running attack

We have seen how an attacker could benefit from unintentional token transfers using a front-running attack. To prevent front-running attacks, there are some techniques we can follow.

You can use the following implementation of the `approve()` function, which ensures that, before updating the value, it should be set to zero. This completely prevents a front-running attack:

```
function approve(address _spender, uint256 _value) public returns (bool) {
//prevent front-running attack
require(_value == 0 || allowed[msg.sender][_spender] == 0);

allowed[msg.sender][_spender] = _value;
emit Approval(msg.sender, _spender, _value);
return true;
}
```

The preceding code will ensure that the value is set to zero between two `approve()` function calls, for the number of tokens (value) more than zero.

You can also prevent these attacks by using advanced functions such as `increaseApproval()` and `decreaseApproval()`. You can find the definitions of these functions later in this chapter.

The transferFrom function

The `transferFrom()` function transfers the tokens from an owner's account to the receiver account, but only if the transaction initiator has sufficient allowance that has been previously approved by the owner to the transaction initiator. To transfer the tokens using the `transferFrom()` function, approver must have called the `approve()` function prior. As per the standard, the `transferFrom()` function must fire the `Transfer` event upon the successful execution and transfer of tokens. The transfer of 0 (zero) value must also be treated as a valid transfer and should fire the `Transfer` event.

The ERC20 standard API is as follows:

```
function transferFrom(address from, address to, uint256 value) external
returns (bool success);
```

The `transferFrom()` function takes three arguments:

- `from`: An `address` type argument to deduct the tokens from the `from` Ethereum account
- `to`: An `address` type argument to transfer the tokens to the `to` Ethereum account
- `value`: A `uint256` type argument to define the number of tokens to deduct from `from` account and send to `to` account.

This function returns a `bool` value representing whether the transferal of the token is executed successfully or not. Standard implementation of the `transferFrom()` function is shown in the following code:

```
function transferFrom(
    address _from,
    address _to,
    uint256 _value
)
    public
    returns (bool)
{
    require(_value <= balances[_from]);
    require(_value <= allowed[_from][msg.sender]);
    require(_to != address(0));

    balances[_from] = balances[_from].sub(_value);
    balances[_to] = balances[_to].add(_value);
    allowed[_from][msg.sender] = allowed[_from][msg.sender].sub(_value);
    emit Transfer(_from, _to, _value);
    return true;
}
```

As per the preceding implementation, you can see that it checks that the destination address should not be `address(0)`. It also verifies that the `_value` is less than or equal to the balance of `msg.sender`.

Let's understand the `transferFrom()` function by working with an example—let's assume that the balances of the accounts are as follows:

Account	TKN token balance
Alice	1,000
Bob	0
Charlie	10

Alice and Bob are both friends. Alice approved 20 TKN tokens for Bob by calling the `approve()` function as the first transaction. Bob is now allowed to withdraw 20 TKN from Alice's account. Now, Bob wants to transfer 5 TKN to Charlie, but he does not own any token. However, he called the `transferFrom()` function to take the 5 TKN from Alice's account and transfer it to Charlie.

The following are the transactions Alice and Bob will perform in order. You can also see how the `allowed` mapping is affected by the transactions:

Tx initiator	Function call	Value before Tx	Value after Tx
Alice	`approve(Bob, 20)`	`allowed[Alice][Bob]=0`	`allowed[Alice][Bob]=20`
Bob	`transferFrom(Alice, Charlie, 5)`	`allowed[Alice][Bob]=20`	`allowed[Alice][Bob]=15`

After these two transactions, the balances of the accounts will be as follows:

Account	TKN token balance
Alice	995
Bob	0
Charlie	15

However, Bob is still allowed to withdraw 15 TKN from Alice's account. Bob can also transfer the tokens to himself by calling the function, like following:

```
transferFrom(Alice, Bob, 15)
```

Two-step process for contracts

If you are calling the `transferFrom()` function from within a Solidity contract, it is recommended to enclose the call with `require()` to ensure that the token transfer executed successfully and, in case of any transfer failure, the transaction should revert. It is defined as follows:

```
require(ERC20.transferFrom(from, to, value));
```

As we have seen, for the `transferFrom()` function to work, the `approve()` function must be called to approve the allowances. It's the same with the Solidity contract. If you have the preceding statement in your Solidity contract, the address of which, let's say, is `0xContractAddress`, the user must have called the `approve(0xContractAddress, amountOfTokens)` function for the contract address to give approval to the contract address. Now, the contract code can call `transferFrom()` and take or withdraw sufficient token balance from the user's account.

The preceding process is used to pay for some services in ERC20 tokens to a contract. For example, you want to subscribe to some service or become a member; to pay for those services or the membership fee, you can pay using some ERC20 tokens.

The allowance function

The `allowance()` function returns the token amount remaining, which the spender is currently allowed to withdraw from the owner's account. This function returns the remaining balance of tokens from the `allowed` mapping. The `allowed` mapping is updated when `approve()`, `transferFrom()`, and advanced functions such as `increaseApproval()` or `decreaseApproval()` are present and executed.

The following is the API for the `allowance()` function:

```
function allowance(address owner, address spender) external view returns
(uint256 remaining);
```

The `allowance()` function takes two arguments:

- `owner`: The address of the `owner` of the token who previously called the `approve()` function
- `spender`: The address of `spender`, who is allowed to withdraw tokens from the owner's address

The function returns the remaining balance of the approved number of tokens as `uint256`.

The implementation of the `allowance()` function reads the state from the `allowed` mapping, as follows:

```
function allowance(address _owner, address _spender) public view returns
(uint256) {
    return allowed[_owner][_spender];
}
```

The balanceOf function

The `balanceOf()` function returns the token balance of the given address. The balances of each account are maintained in a mapping. This function just returns the `uint256` balance data from the mapping.

The following is the API of the `balanceOf()` function:

```
function balanceOf(address who) external view returns (uint256 balance);
```

The `balanceOf()` function takes a single argument:

- `who`: The address whose token balances you want to read

The function returns the token balance as a `uint256` type. The following is the function implementation, where it reads the data from the `balances` mapping and returns the token balance:

```
function balanceOf(address _owner) public view returns (uint256) {
    return balances[_owner];
}
```

The totalSupply function

The `totalSupply()` function returns the total supply of the tokens. This means that the sum total of token balances of all of the token holders must match the total supply. The state of the total supply is maintained in a state variable, which gets increased or decreased upon minting new tokens (creation of new tokens) or burning (destroying existing tokens). The `transfer()` or `transferFrom()` functions do not update the total supply of the token.

The following is the API of the `totalSupply()` function:

```
function totalSupply() external view returns (uint256);
```

The function call returns the total supply of the token as `uint256`. The following is the code that just returns the value of the `totalSupply_` variable:

```
function totalSupply() public view returns (uint256) {
    return totalSupply_;
}
```

However, it is also possible to add a `totalSupply` state variable that has `public` visibility. For this, the Solidity compiler will auto-generate the `view` function.

Events

As per the ERC20 standard, there are only two events that need to be logged when a certain function of the contract is called.

The events will only be logged when the transaction is successful. For all failed transactions, these events will not be triggered and logged.

As we saw in the previous chapters, Solidity events can also be filtered. On the client applications, you can subscribe to the topics to listen for the events when they are logged on the Ethereum blockchain. Once these events are triggered, on the client application, you get a notification. Based on these notifications, you can perform your actions. These notifications are push notifications from the Ethereum blockchain (the node you are connected with) to your client application.

Now, let's look into the `Transfer` event.

The Transfer event

This event must be triggered when the tokens are transferred from one account to another account. Even if the number of tokens is zero, the event must trigger. When new tokens are created or minted, the event should use `0x0` as the `from` address.

The `Transfer` event in the Solidity contract must be defined as follows:

```
event Transfer(address indexed from, address indexed to, uint256 value);
```

The event takes the following arguments:

- `from`: This is the address from which the tokens are sent. This parameter is indexed.
- `to`: This is the address that receives the tokens. This parameter is indexed.
- `value`: This is the number of tokens (as `uint256`) transferred from the `from` address to the `to` address.

The Approval event

The `Approval` event must be triggered on any successful call to the `approve()` function, even if the approval call is made for a zero amount. The zero value approval is also valid and can be given to an address to revoke approvals that had been given to someone prior to this.

In the new updated ERC20 implementations of OpenZeppelin, the `Approval` event is triggered even when the allowed balance is changed while executing the `transferFrom()` function. We will look into the new implementation of ERC20 in `Chapter 9`, *Deep Dive into the OpenZeppelin Library*.

The `Approval` event in the Solidity contract must be defined as follows:

```
event Approval(address indexed owner, address indexed spender, uint256 value);
```

The event takes the following arguments:

- `owner`: This is the owner address that has given approval. This is an indexed parameter.
- `spender`: This is the spender address that receives approval. This is an indexed parameter.
- `value`: This is the number of tokens (as `uint256`) `owner` has approved for `spender` to spend on their behalf.

Optional functions

In the ERC20 standard specification, there are a few functions present, but those are marked as optional functions. It is up to the developer to decide whether to implement these functions or not, as per their requirements.

The names of these functions are `name`, `symbol`, and `decimals`. In almost all of the ERC20 contracts, if they need these functions, they just define the `public` variable using these names as the state variable of the contract. This is because, as per Solidity, the getter methods are created automatically by the compiler for any `public` state variable. Hence, if you are creating an ERC20 token, you can simply add these state variables and assign the values according to your token metadata, as shown in the following code:

```
string public name = "My Test Token";
string public symbol = "TKN";
uint8 public decimals = 18;
```

The name function

The `name()` function should return the token's full name, for example, `My Test Token`. The `name()` function is defined in the contract as follows or using the `public` state variable:

```
function name() public view returns (string)
```

The function's visibility can be `public` or `external`. It must be a `view` function and should not be allowed to modify the state. The function must return a `string` type value.

The symbol function

The `symbol()` function should return the token's symbol, for example, `TKN`. The `symbol()` function is defined in the contract as follows or using the `public` state variable:

```
function symbol() public view returns (string)
```

The function's visibility can be `public` or `external`. It must be a `view` function and should not be allowed to modify the state. The function must return a `string` type value.

The decimals function

The `decimals()` function should return the fungibility of the token in the number of decimals. For example, 18 means to divide the token amount by 10^{18} to get its whole value. The `decimals()` function is defined in the contract as follows or using the `public` state variable:

```
function decimals() public view returns (uint8)
```

The function's visibility can be `public` or `external`. It must be a `view` function and should not be allowed to modify the state. The function must return a `uint8` type value.

Advanced functions

There are some advanced functions that were recently added in the ERC20 token implementations. However, these functions are not part of the ERC20 standard APIs. These functions are just added to improve usage and reduce security issues and attacks. If you are writing decentralized or centralized exchanges, these functions should not be taken as part of the ERC20 standard as these functions may not be found in all of the ERC20 standard token implementations.

In the new OpenZeppelin implementation of ERC20 contracts, there are more functions such as _mint(), _burn(), and _burnFrom() that were also added. We will look into those functions in detail in Chapter 9, *Deep Dive into the OpenZeppelin Library*.

As we have seen when we talked about the approve() function, it's possible to attack the function using front-running techniques. By using the increaseApproval() and decreaseApproval() functions to increase or decrease the approval for the number of tokens, the attacks can be eliminated. For example, after calling the approve() function once, if you want to change it without being attacked using front-running, you can use the increaseApproval() or decreaseApproval() functions.

The increaseApproval function

This function is called by the approver to increase the approved allowed amount of the spender. This just increases the number of previously approved tokens by adding _addedValue. The function also triggers the Approval event to notify the updated approved balance:

```
function increaseApproval(
    address _spender,
    uint256 _addedValue
)
    public
    returns (bool)
{
    allowed[msg.sender][_spender] = (
    allowed[msg.sender][_spender].add(_addedValue));
    emit Approval(msg.sender, _spender, allowed[msg.sender][_spender]);
    return true;
}
```

This function takes the following arguments:

- _spender: The spender's address whose approved amount is to be increased
- _addedValue: The number of tokens with which to increase the previous allowance

For example, an owner has already given approvals for 100 TKN tokens to the approved person. The owner can call increaseApproval() function with following parameters:

- _spender: The address of the approved person
- _addedValue: 50 (the number of TKN tokens to be given approval for)

At the end of the preceding function call, the approved person would have approvals for 150 TKN tokens.

The decreaseApproval function

This function is called by the approver to decrease the approved allowed amount of the spender. This just decreases the number of previously approved tokens by subtracting _subtractedValue. This function can also be used to completely revoke the approvals. The function also triggers the Approval event to notify the updated approved balance:

```
function decreaseApproval(
    address _spender,
    uint256 _subtractedValue
)
    public
    returns (bool)
{
    uint256 oldValue = allowed[msg.sender][_spender];
    if (_subtractedValue >= oldValue) {
        allowed[msg.sender][_spender] = 0;
    } else {
        allowed[msg.sender][_spender] = oldValue.sub(_subtractedValue);
    }
    emit Approval(msg.sender, _spender, allowed[msg.sender][_spender]);
    return true;
}
```

This function takes the following arguments:

- _spender: The spender's address whose approved amount is to be decreased
- _subtractedValue: The number of tokens to decrease from the previous allowance balance

For example, an owner has already given approvals for 100 TKN tokens to the approved person. The owner can call the decreaseApproval() function with the following parameters:

- _spender: The address of the approved person
- _subtractedValue: 20 (the number of TKN tokens to be reduced)

At the end of the preceding function call, the approved person would have approvals for 80 TKN tokens.

Summary

In this chapter, we understood the most famous Ethereum token standard—ERC20. We looked into the different aspects of the standard, the function definition, and its usage, which a developer must know about. We discussed the approve() function, how front-running attacks are possible using that function, and how to protect your code against it. We also look at some of the advanced functions of the ERC20 implementations, which are not part of the standard. Many other token standards have been published; however, ERC20 is still used for basic token implementation. According to your contract architecture and requirements, you can use other token standards, but you need to ensure that the token standard at least follows ERC20 APIs as well.

In the next chapter, we will learn about the ERC721 non-fungible tokens standard, which is another popular token standard that's used for games and non-fungible items like digital crypto collectible cards.

Questions

1. Is **ether** (symbol: **ETH**) ERC20-compliant?
2. How can you pay using ERC20 tokens for a service?
3. Are there more advanced token standards?
4. Is it possible to transfer tokens in bulk to multiple addresses?
5. What can we do once the token is locked into a contract?

ERC721 Non-Fungible Token Standard

8

We looked into the ERC20 token standard in `Chapter 7`, *ERC20 Token Standard*. ERC20 is a fungible token standard; however, `ERC721` is a **Non-Fungible Token (NFT)** standard. This standard is used in many cases where you want to transfer a whole item that cannot be broken into multiple parts, for example, a house deed or collectible cards. These items are non-fungible as these represent a whole entity and cannot be sub-divided into multiple pieces. For example, a collectible card in itself is a whole entity and has some economic value.

In this chapter, we will have a look into the `ERC721` standard functions and its implementation in detail. We will also look into optional `ERC721` metadata and enumeration-related interfaces and implementation.

We will cover the following topics in this chapter:

- Overview of the `ERC721` non-fungible token standard
- `ERC721` interface functions
- Understanding `ERC721` implementation
- Understanding the `ERC721` receiver
- Optional `ERC721` metadata functions
- Optional `ERC721` enumeration functions

Technical requirements

The code discussed in this chapter is present on GitHub at `https://github.com/PacktPublishing/Mastering-Blockchain-Programming-with-Solidity/tree/master/Chapter08`.

In this chapter, we will discuss the `ERC721` non-fungible token standard and its implementation only. Hence, there is no technical requirements for this chapter. You just need to refer to the code present at the GitHub location.

Overview of the ERC721 NFT standard

In `Chapter 7`, *ERC20 Token Standard*, we looked into the ERC20 token standard, which is the standard that's mostly used for minting and transferring the tokens. However, the ERC20 token standard has a state variable, `decimals`, and by using that variable, each token of the token contract can be fungible to the defined number of decimal places. For example, an ERC20 token with 18 decimal places would enable each of its minted tokens to be fungible up to 18 decimal places. The Solidity language does not support decimal or floating data types; hence, one full unit of a token will have 18 zeros followed by 1, which we can also refer to as 1^{18} or $10*18$. These tokens are also identical to each other, which means that one token is equal to another token. For example, if there are two people who both have 1 **OmiseGo** (Symbol: **OMG** ERC20 token) in their wallets, they both have identical tokens that have the same economic value present in their wallets. Because of this divisible and identical property, the ERC20 tokens are mostly used for financial applications.

On the other hand, the `ERC721` standard is a NFT standard, in which an NFT is non-divisible and non-fungible. It means an NFT in itself is a complete token that cannot be further sub-divided. Each of the NFTs has a unique ID to represent its uniqueness among all of the `ERC721` tokens minted from the same contract. Here, uniqueness also means that an `ERC721` token is not equal to another `ERC721` token. As `ERC721` tokens are non-fungible and unique, they define the ownership of a property registry or deed. If you own an `ERC721` NFT that represents the ownership of real estate, it means that you own a piece of real estate that cannot be subdivided. NFTs can be used to represent the ownership of a digital or physical asset; for example, CryptoKitties is an Ethereum game in which each digital collectible kitten is represented with a unique NTF token. NFTs can also be used to represent a piece of digital art, that is unique and non-fungible.

Here are the basic differences between the ERC20 token standard and the ERC721 NFT standard:

Property	The ERC20 token standard	The ERC721 NFT standard
Fungibility	Tokens are fungible up to the decimal places defined in the contract.	Tokens are non-fungible. Each token itself represents 1 token, which cannot be sub-divided.
Ownership	Token ownership is not directly linked to an account, and only the token balances are tracked in the contract.	Each token's ownership is linked to an individual account address.
Uniqueness	One token is similar in economic value to another token from the same token contract. There is no difference in each token.	Each token is different than each other token present in the same contract and may have different economic value.

The ERC721 NFT standard primarily got hyped when it was used in one of the famous Ethereum on-chain game called **CryptoKitties** back in 2017. In the game, there are cats; each one is different from the other and you can buy, sell, and breed cats. Each cat is represented as an ERC721 token, which is unique. Each cat has specific and unique properties such as skin color, eye color, and hair color. You can also try this game online at https://www.cryptokitties.co/.

As we have learned about ERC721 NFT standard, we can now look into its standard interface, which provides different functions with which one can interact with the ERC721 token implementation.

The ERC721 NFT standard API interface

The specification of the ERC721 standard interface that we are discussing in this section is taken from the EIP standard document at https://github.com/ethereum/EIPs/blob/master/EIPS/eip-721.md.

The ERC721 NFT standard interface is defined as follows:

```
interface ERC721 is ERC165 {
    event Transfer(address indexed _from, address indexed _to,
                   uint256 indexed _tokenId);

    event Approval(address indexed _owner, address indexed _approved,
                   uint256 indexed _tokenId);


    event ApprovalForAll(address indexed _owner, address indexed _operator,
                         bool _approved);
```

```
    function balanceOf(address _owner) external view returns (uint256);

    function ownerOf(uint256 _tokenId) external view returns (address);

    function safeTransferFrom(address _from, address _to,
                             uint256 _tokenId, bytes data)
                             external payable;

    function safeTransferFrom(address _from, address _to,
                             uint256 _tokenId) external payable;

    function transferFrom(address _from, address _to,
                         uint256 _tokenId) external payable;

    function approve(address _approved, uint256 _tokenId) external payable;

    function setApprovalForAll(address _operator, bool _approved) external;

    function getApproved(uint256 _tokenId) external view returns (address);

    function isApprovedForAll(address _owner, address _operator)
        external view returns (bool);
}
```

The preceding code defines the basic functions and events that an ERC721 NFT implementation must follow. We will discuss each of the functions and events in the upcoming sections.

As you can see in the code, the ERC721 interface also inherits from the ERC165 standard. The ERC165 standard is known as the **Standard Interface Detection**, using which we can publish and detect all interfaces a smart contract implements.

Please note that the preceding inheritance definition has the name ERC721. However, in OpenZeppelin's source code, the interface name is IERC721 and its implementation contract name is ERC721; hence, the reader should not be confused by the similarities in name.

Understanding the ERC721 implementation

As we have seen the ERC721 standard interface, now let's look at the ERC721 basic implementation. Here, for the implementation, we are going to use the code provided in Zeppelin's openzeppelin-solidity 2.1.1 release. However, developers must keep checking for new releases of the OpenZeppelin project to ensure that any bug fixes or gas improvements are updated in the latest code. You can always check for latest releases and the included bug fixes of OpenZeppelin on https://github.com/OpenZeppelin/openzeppelin-solidity/releases.

We will look into the following in this section:

- State variables used in the ERC721 implementation
- The different functions of ERC721 and their implementation
- Internal functions provided by OpenZeppelin's ERC20 implementation
- Events present in the ERC721 implementation and their usage

The following are the source code files present at our GitHub repository that we will discuss in this section:

- Chapter08/contracts/IERC721.sol: The ERC721 interface
- Chapter08/contracts/ERC721.sol: The ERC721 implementation
- Chapter08/contracts/introspection/IERC165.sol: The ERC165 interface
- Chapter08/contracts/introspection/ERC165.sol: The ERC165 implementation

ERC721 inherits from IERC721 and ERC165

The ERC721.sol contract contains the basic implementation of the ERC721 NFT Standard. The ERC721 contract inherits from other contracts: ERC165 and IERC721. The inheritance is defined in the contract as follows:

```
contract ERC721 is ERC165, IERC721 {
```

The ERC165 and IERC721 contracts further inherit IERC165, as shown in the following diagram:

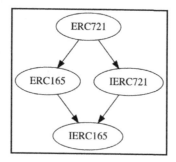

Inheritance of the ERC721 contract generated using the surya tool

The ERC721 contract contains the basic implementation of ERC721 standard. The contract imports from IERC721 interface and provides the definition to each function. Furthermore, the ERC721 contract implements ERC165 features to let the function caller know which of the function calls are supported by the ERC721 contract.

Now, let's understand the features that are inherited from ERC165 and IERC721 into the ERC721 contract.

The preceding inheritance graph is generated using the surya tool. You can refer to Chapter 6, *Taking Advantage of Code Quality Tools* section *Generating an inheritance graph* and learn to generate inheritance graph.

ERC721 inherits from ERC165

The ERC721 contract inherits from the ERC165 contract, which contains its implementation. ERC165 also inherits from IERC165, which is the interface definition for ERC165 standard interface detection. The ERC165 contract inheritance is defined as follows:

```
contract ERC165 is IERC165 {
```

As we have discussed, ERC165 is used for standard interface detection, meaning that it allows the client application/contract to know whether a particular function is available to call on the contract. The contract defines the following function:

```
function supportsInterface(bytes4 interfaceId) external view returns (bool)
```

Using this function, we can pass the first four bytes of the function signature that the caller wants to check whether the target contract supports it or not. If the contract has the function present in the contract and it is supported, the preceding function call would return `true`; otherwise, it would return `false`.

If you are using ERC165 in your project, you should ensure that it is used for `public` and `external` functions only, which are supported by your contract. As the `private` and `internal` functions are not exposed outside of the contract, those functions should not be registered in your ERC165 contract.

ERC721 inherits from IERC721

As we have seen in the previous inheritance graph, the ERC721 contract inherits from IERC721. The IERC721 is the interface definition present in the OpenZeppelin library. The contract defines all basic events and functions that the ERC721 NFT standard constitutes. The IERC721 interface also inherits from IERC165, which also contains the interface definition for ERC165 standard interface detection. The following is the contract inheritance of IERC721.sol:

```
contract IERC721 is IERC165 {
```

This interface is defined as the preceding preset in the OpenZeppelin source code. We looked into the ERC721 interface graph and the corresponding contracts it inherits from; now, let's look into the different state variables present in the ERC721 implementation.

Understanding ERC721 state variables

The ERC721 contract definition of OpenZeppelin uses some state variables to keep track of the NFTs and their ownership and approvals. It also has some constants defined that are used for the ERC165 standard, so that the client can detect the functions supported by the contract.

Let's look at each of the state variables defined in the contract.

Token owner mapping kept in _tokenOwner

The mapping is used to find the owner of a given NFT ID. In the code, it's defined as follows:

```
mapping (uint256 => address) private _tokenOwner;

function ownerOf(uint256 tokenId) public view returns (address) {
    address owner = _tokenOwner[tokenId];
    require(owner != address(0));
    return owner;
}
```

_tokenOwner is a mapping from uint256 to address, which stores the mapping for an NFT ID to the account address of the owner.

The mapping variable is private, and the ownerOf() accessor view function fetches the data from this state variable. The ownerOf() function finds the current owner of the given tokenId.

Approved address mapping kept in _tokenApprovals

An owner of an ERC721 token can give approval to another address. Once approval is given, another address can transfer the approved ERC721 token from the owner's wallet. Let's understand how these approvals are maintained in the contract.

The _tokenApprovals mapping is used to find the approver's address for a given NFT ID. _tokenApprovals is a mapping from uint256 to address, which stores the mapping for the NFT ID to the approved spender's address. In the code, it's defined as follows:

```
mapping (uint256 => address) private _tokenApprovals;

function getApproved(uint256 tokenId) public view returns (address) {
    require(_exists(tokenId));
    return _tokenApprovals[tokenId];
}
```

The mapping variable is private, and the getApproved() accessor view function fetches the read-only data from this state variable. Using the getApproved() function, anyone can provide the tokenId and find its approved address. Further down in this chapter, we have a discussion about the getApproved() function in detail. The approve() function is used by the owner to give approvals for a given token ID to another address.

The number of tokens per owner kept in _ownedTokensCount

The mapping is used to find the number of tokens an address owns. _ownedTokensCount is a mapping from `address` to `uint256`, which stores the mapping for a wallet address and the corresponding number of NFTs it owns. In the code, it's defined as follows:

```
mapping (address => uint256) private _ownedTokensCount;
```

The `mapping` variable is `private`, and the `balanceOf()` accessor function reads the data from this state variable.

Operator approvals kept in _operatorApprovals

An operator is a potential, trusted third party that can be added by an owner who owns one or more NFT tokens.

If an owner of one or more NFT tokens gives approval to an operator (another address), then the operator is allowed to withdraw any of the owner's tokens from the owner's address, and he is allowed to send it to another address or to themselves.

Just to simplify, an operator has approvals for all the NFT tokens present in the owner's account.

The `mapping` of a `mapping` is used to find the operator's addresses that are approved by a given owner's address:

```
mapping (address => mapping (address => bool)) private _operatorApprovals;

function setApprovalForAll(address to, bool approved) public {
    require(to != msg.sender);
    _operatorApprovals[msg.sender][to] = approved;
    emit ApprovalForAll(msg.sender, to, approved);
}

function isApprovedForAll(address owner, address operator) public view
returns (bool) {
 return _operatorApprovals[owner][operator];
}
```

_operatorApprovals is a mapping of a mapping defined as in the preceding code, where the approver's address is used as a key in the first mapping and the operator's address is used as a key in the second mapping. The Boolean value stored in the second mapping signifies whether the approval is given or not to the operator address.

The mapping variable is private, the isApprovedForAll() accessor function reads the data from the state variable, and the setApprovalForAll() function is used by the owner to give approvals to an operator. Once the approval is given to an operator, they can transfer any ERC721 token from the owner's account. As it's a mapping of a mapping, the owner can give approvals to many addresses, and can revoke the access as well. An owner can give operator approval to one or more addresses.

The ERC165 interface code for the ERC721, _INTERFACE_ID_ERC721

As discussed in the previous section, where we looked into the ERC165 interface and implementation, the API requires a four-byte method signature, which is supported by the contracts. In the ERC721 implementation, the _INTERFACE_ID_ERC721 constant state variable is assigned a hardcoded value, 0x80ac58cd, which is generated by doing an **Exclusive OR (XOR)** operation of each function signature present in the IERC721 interface. In Solidity, the ^ (caret) operator is used to perform bitwise XOR operations. The constant value generation is shown in the following code:

```
0x80ac58cd ==
    bytes4(keccak256('balanceOf(address)')) ^
    bytes4(keccak256('ownerOf(uint256)')) ^
    bytes4(keccak256('approve(address,uint256)')) ^
    bytes4(keccak256('getApproved(uint256)')) ^
    bytes4(keccak256('setApprovalForAll(address,bool)')) ^
    bytes4(keccak256('isApprovedForAll(address,address)')) ^
    bytes4(keccak256('transferFrom(address,address,uint256)')) ^
    bytes4(keccak256('safeTransferFrom(address,address,uint256)')) ^
    bytes4(keccak256('safeTransferFrom(address,address,uint256,bytes)'))
```

And the bytes4 value generation is hardcoded in the ERC721 contract, as shown in the following:

```
bytes4 private constant _INTERFACE_ID_ERC721 = 0x80ac58cd;
```

If any of the contracts support ERC165 and have the 0x80ac58cd value registered as a supported interface, it means that the contract supports the standard API functions of the ERC721 NFT standard. We can check this by calling the supportsInterface() function on the contract and providing 0x80ac58cd as the function argument. If the function returns true, then that contract supports all ERC721 functions.

The ERC165 interface code for the ERC721Receiver, _ERC721_RECEIVED

The bytes4 method signature code is used to identify that the onERC721Received() function is supported by the contract. This function is present in IERC721Receiver, which is used to receive the token callback call on the contract. We will discuss the ERC721Receiver contract in the next section in this chapter:

```
0x150b7a02 ==
bytes4(keccak256("onERC721Received(address,address,uint256,bytes)"))
```

The bytes4 value for the variable is calculated as shown in the preceding code. The signature of the onERC721Received() function is first passed in the keccak256() built-in Solidity function. The output hash of the function is further passed to the bytes4() built-in Solidity function that only takes the first 4 bytes of the hash.

Here, the keccak256() built-in Solidity function is an alias to the sha3() function. The function generates an Ethereum SHA256 hash of the given parameter:

```
bytes4 private constant _ERC721_RECEIVED = 0x150b7a02;
```

The value represents the function signature of the onERC721Received() function. The function signature is hardcoded in the ERC721 contract.

We understood why and how the particular state variable is defined, and that it stores values present in the ERC721 contract. Now, let's understand each of its functions.

The constructor of ERC721

In the ERC721 implementation, the constructor is defined as follows:

```
constructor () public {
    // register the supported interfaces to conform to ERC721 via ERC165
    _registerInterface(_INTERFACE_ID_ERC721);
}
```

As you can see in the preceding code, it registers the `bytes4` signature of the `ERC721` interface functions using the `ERC165` internal function, `_registerInterface()`. By registering this, the `ERC721` contract allows any other contract to call `supportsInterface()` and find the list of functions supported by the deployed instance of the `ERC721` contract.

The balanceOf function

As we have seen in Chapter 7, *ERC20 Token Standard*, the ERC20 implementation also has a `balanceOf()` function to check the token balance of a particular account. Here, in the case of the `ERC721` implementation, the function returns the number of NFT tokens that the provided address owns:

```
function balanceOf(address owner) public view returns (uint256) {
    require(owner != address(0));
    return _ownedTokensCount[owner];
}
```

As you can see in the preceding code that if the owner address provided in the function argument is `address(0)`, the function call will fail. Here, `address(0)` is also represented as `address(0x0)` or `0x0000000000000000000000000000000000000000`. As in the `ERC721` implementation, it is not possible to send tokens to `address(0)`; hence, reading the token balance of that address is also not valid.

Anyone is allowed to call this function, as it is a `public` and `view` function, to read the value for off-chain calculation as well.

The ownerOf function

We discussed that, in an `ERC721` contract, each NFT is unique and has a different economic value. Each NFT has an associated owner. However, an owner can have multiple NFTs. Hence, between an owner and NFT, there is a one-to-many relationship.

The `ownerOf()` function finds the provided `tokenId` in the `_tokenOwner` map and returns the `address` of its `owner`:

```
function ownerOf(uint256 tokenId) public view returns (address) {
    address owner = _tokenOwner[tokenId];
    require(owner != address(0));
    return owner;
}
```

As you can see in the preceding code that in case the `tokenId` provided does not exist in the `mapping`, the `owner` local variable will be assigned `address(0)` (the default value for the `address` datatype) and the call to the function will fail.

Anyone is allowed to call this function, as it is a `public` and `view` function, to read the value for off-chain calculation as well.

The approve function

Using the `approve()` function, the owner of an NFT can give approval to another account, so that the approved account can transfer the approved NFT from the owner's account.

As you can see in the following code, the function takes two arguments:

- `to`: The address to which approval is to be given. Once approval is given, the `to` address is allowed to do anything with `tokenId`.
- `tokenId`: The NFT `tokenId` for which approval is given.

The code for the `approve` function is as follows:

```
function approve(address to, uint256 tokenId) public {
    address owner = ownerOf(tokenId);
    require(to != owner);
    require(msg.sender == owner || isApprovedForAll(owner, msg.sender));

    _tokenApprovals[tokenId] = to;
    emit Approval(owner, to, tokenId);
}
```

As you can see in the preceding code, the `approve()` function can also be called by the operator (who has rights for the `owner`'s tokens). The operator can further give approvals to another account. There are two `require` checks, the one `require(to != owner)` ensures that the operator cannot give further approvals to the token `owner`. The second `require` ensures that the sender is either the `owner` or has operator rights. After these checks are verified, token approval is given to the `to` address for the given `tokenId`. Finally, it emits an `Approval` event along with the approval data.

One thing to note is that the same functionality is not possible in the ERC20 implementation. The approver cannot give further approvals to another account.

The getApproved function

The getApproved() function returns the address that has the approval for the given tokenId. As you can see, the function takes tokenId as a function argument for which you want to know the address of the approved account:

```
function getApproved(uint256 tokenId) public view returns (address) {
    require(_exists(tokenId));
    return _tokenApprovals[tokenId];
}
```

As you can see in the preceding code that the function calls an internal _exists() function to ensure that the given tokenId of an NFT exists, and that it has been assigned to a non address(0) owner, otherwise, the function call fails. After this check, it fetches and returns the token owner's address.

The setApprovalForAll function

The setApprovalForAll() function is used to assign or revoke the full approval rights to the given operator. The caller of the function (msg.sender) is the approver.

The function takes the following arguments:

- to: This is the address of the operator to whom the approval rights should be given or revoked from the approver.
- approved: This is the Boolean representing the approval to be given or revoked, and the argument is based on the following:
    - true: The approval will be given to the operator.
    - false: The approval will be revoked from the operator.

Let's look at the setApprovalForAll() function code:

```
function setApprovalForAll(address to, bool approved) public {
    require(to != msg.sender);
    _operatorApprovals[msg.sender][to] = approved;
    emit ApprovalForAll(msg.sender, to, approved);
}
```

In the preceding code, you can see that the function has a require() statement, which ensures that the caller cannot give operator approvals to himself. Further, this function maintains the approval rights in the _operatorApprovals mapping. It also emits the ApprovalForAll event.

The isApprovedForAll function

The isApprovedForAll() function is used to check that the given operator has operator rights on the given owner's tokens.

This function takes the following arguments:

- owner: The address of owner of the tokens who has given full approvals before
- operator: The address of operator that has received the approval from owner before

This function returns a Boolean value. The following are the return values:

- true: The owner has previously given full approvals to the operator
- false: The owner has not given approval to the operator

Let's look at the isApprovedForAll() function code:

```
function isApprovedForAll(address owner, address operator) public view
returns (bool) {
    return _operatorApprovals[owner][operator];
}
```

As you can see in the preceding code that the function has the public visibility and a view modifier; hence, you would be able to call this function off-chain as well. It fetches the read-only data from the _operatorApprovals mapping.

The transferFrom function

The transferFrom() function is a public function, used to transfer the given tokenId from the owner's account to the receiver's account. For this function to work, the approval must have been given previously by the owner to the address of the caller of this function.

This function takes the following arguments:

- from: The address of the owner who currently owns the given NFT
- to: The address of the receiver of the NFT
- tokenId: The tokenId of the NFT that is to be transferred to the to address

Let's look at the transferFrom() function code:

```
function transferFrom(address from, address to, uint256 tokenId) public {
    require(_isApprovedOrOwner(msg.sender, tokenId));

    _transferFrom(from, to, tokenId);
}
```

As you can see in the preceding code that the function first validates that the msg.sender (function caller) has the required approvals to transfer the given tokenId. Once it is verified, the internal call to the _transferFrom() function transfers the token from the owner's account to the recipient's account.

The safeTransferFrom function

The safeTransferFrom() function is a public function that is used to safely transfer the NFT from the owner's account to the recipient account. Here, safely transfer means that, when a recipient is a contract, you need to ensure that the recipient contract has callbacks registered to let the receiver contract get a notification when ERC721 token transfer is successful.

This function takes the following arguments:

- from: The address of the current owner of the NFT
- to: The address of the recipient who will receive the NFT
- tokenId: tokenId of the NFT that is to be transferred
- _data: The bytes data that is to be forwarded to the ERC721Receiver contract, if the to address (recipient) is a contract

Let's look at the safeTransferFrom() function code:

```
function safeTransferFrom(address from, address to, uint256 tokenId, bytes
memory _data) public {
    transferFrom(from, to, tokenId);
    require(_checkOnERC721Received(from, to, tokenId, _data));
}
```

As you can see in the preceding code that the function calls the transferFrom() function to transfer the NFT to the recipient address. Then, it makes a call to the _checkOnERC721Received() internal function, which further calls the callback functions (the onERC721Received() function on the contract receiving the token) in case the recipient of the NFT is a contract (not an **Externally Owned Account (EOA)**).

This function is useful when you want to pass on some further contract function calls to be made from the receiver's contract that received the given token. You can pass on the function bytes data into the `safeTransferFrom()` function in the _data argument. When this _data parameter is not empty, the further function call will be initiated from the receiver's `onERC721Received()` function. When this _data parameter is an empty string, no further function calls will be made from the receiver's contract.

Another safeTransferFrom function

This `safeTransferFrom()` function is mostly the same as the previous one, with one big difference—this is the overloaded function, having different function arguments. This function is used when you do not want to make further function calls from the recipient's callback function call.

The function takes the following arguments:

- `from`: The address of the current owner of the NFT
- `to`: The address of the recipient who will receive the NFT
- `tokenId`: The `tokenId` of the NFT that is to be transferred

Let's look at the `safeTransferFrom()` function code:

```
function safeTransferFrom(address from, address to, uint256 tokenId) public
{
    safeTransferFrom(from, to, tokenId, "");
}
```

As you can see in the preceding code, it simply makes a call to the other `safeTransferFrom()` function, which takes four arguments. However, the _data parameter is passed as empty string. So, this function should be used in the following cases:

- When the recipient of the NFT is an EOA account (not a contract account)
- When the recipient of the NFT is a contract account (not an EOA account), and you do not want to call any further functions in the `ERC721Receiver` contract

The _exists internal function

`_exists()` is an `internal` function that is used to check that the given `tokenId` exists in the contract.

Let's look at the _exists() function code:

```
function _exists(uint256 tokenId) internal view returns (bool) {
  address owner = _tokenOwner[tokenId];
    return owner != address(0);
}
```

As you can see in the preceding code, the provided tokenId is checked in the _tokenOwner mapping. If the owner is address(0), the function returns false; otherwise, it returns true.

The _isApprovedOrOwner internal function

_isApprovedOrOwner() is an internal function, which is used to check whether the given address of spender is approved or an owner of the given tokenId.

This function takes two arguments:

- spender: The address of the spender who is going to spend the given tokenId
- tokenId: The tokenId of the NFT that the spender wants to transfer

This function returns the Boolean result. If the result is as follows:

- true: The given spender is the owner or an approved address
- false: The given spender is neither the owner, nor the approved address

Let's look at the _isApprovedOrOwner() function code:

```
function _isApprovedOrOwner(address spender, uint256 tokenId) internal view
returns (bool) {
    address owner = ownerOf(tokenId);
    return (spender == owner || getApproved(tokenId) == spender ||
        isApprovedForAll(owner, spender));
}
```

As you can see in the preceding code, the owner of tokenId is fetched and checked against spender, getApproved(), and isApprovedForAll(). Here, it is checking that the spender is either the owner, or was previously approved by the owner or an operator for the given tokenId. If either of the conditions returns true, then spender is approved, or is the owner of the given tokenId.

The _mint internal function

The _mint() internal function is used to mint a new NFT at the given address. As the function is internal, it can only be used from inherited contracts to mint new tokens.

This function takes the following arguments:

- to: The address of the owner for whom the new NFT is minted
- tokenId: The new tokenId for the token that will be minted

Let's look at the _mint() function code:

```
function _mint(address to, uint256 tokenId) internal {
    require(to != address(0));
    require(!_exists(tokenId));

    _tokenOwner[tokenId] = to;
    _ownedTokensCount[to] = _ownedTokensCount[to].add(1);

    emit Transfer(address(0), to, tokenId);
}
```

As you can see in the preceding code, the new tokenId is checked using the _exists() function, which ensures that tokenId must not be present in the contract. Each tokenId is unique in ERC721 contracts, hence you would not be able to create two tokens having the same tokenId. Once the validation check is done, the owner is assigned ownership of this new token. Also, the number of tokens the owner has increased by one, to ensure the correct count of the tokens. At the end, the function also emits a Transfer event, in which address(0) is the from address representing the mint address.

The _burn internal function

_burn() is an internal function, which is used to burn a given NFT from the owner's account.

This function takes two arguments:

- owner: The address of the owner whose token is to be burned
- tokenId: The NFT Id that is to be burned from the owner's account

Let's look at the _burn() function code:

```
function _burn(address owner, uint256 tokenId) internal {
    require(ownerOf(tokenId) == owner);

    _clearApproval(tokenId);

    _ownedTokensCount[owner] = _ownedTokensCount[owner].sub(1);
    _tokenOwner[tokenId] = address(0);

    emit Transfer(owner, address(0), tokenId);
}
```

The code first checks that the provided owner address is correct and is actually the owner of the given tokenId. Then, it clears all of the approval entries from the mappings for the given tokenId. Following that, the owner's token count is reduced with one and the token owner is set to address(0). At the end, the function emits the Transfer event, in which address(0) is used as to address representing the burn address.

This function should be used when you know both the owner address of the token and the tokenId which he owns. In practice, this can be used to expose a function from an inherited contract to allow a token owner to burn his own tokens only. For example, you can implement the following logic in an inherited contract:

```
function burn(uint256 tokenId) public {
    super._burn(msg.sender, tokenId);
}
```

The preceding code passes msg.sender as the owner parameter to the _burn() function. Hence, the function caller can burn only his own tokens.

Another _burn internal function

There are two _burn() internal functions present in the implementation. These are overloaded functions that have the same name; however, they take different arguments.

The function takes the tokenId of an NFT that is to be burned. The _burn() function code is defined as follows:

```
function _burn(uint256 tokenId) internal {
    _burn(ownerOf(tokenId), tokenId);
}
```

The code finds the owner of the token and calls the other overloaded _burn() function. As it is an internal function, this can be used in the inherited contract to burn specific tokens when required.

This function should be used by the inherited contracts when they directly want to burn a token using just its tokenId only. For example, you can implement the following logic in an inherited contract:

```
function process(uint256 tokenId) public onlyOwner {
    //Some pre-validations
    bool isAllowedToBurn = true; //process this boolean in logic
    if( isAllowedToBurn ) {
        super._burn(tokenId);
    }
}
```

As in the preceding example, only after certain validations and when it has been allowed to burn a given token, contract can burn tokens directly without checking their owners.

The _transferFrom internal function

_transferFrom() is an internal function, which is used to transfer the given token from the current owner's account to the receiver's account.

The function takes three arguments:

- from: The address of the owner who currently owns the given token
- to: The address of the receiver or the new owner that will receive the transferred token
- tokenId: The NFT ID of the token that is to be transferred from the from address to the to address

Let's look at the _transferFrom() function code:

```
function _transferFrom(address from, address to, uint256 tokenId) internal
{
    require(ownerOf(tokenId) == from);
    require(to != address(0));

    _clearApproval(tokenId);

    _ownedTokensCount[from] = _ownedTokensCount[from].sub(1);
    _ownedTokensCount[to] = _ownedTokensCount[to].add(1);
```

```
    _tokenOwner[tokenId] = to;

    emit Transfer(from, to, tokenId);
}
```

The function first verifies that the `from` address is the current owner of the token. Then, it clears any approvals related to the token. This is followed by decreasing the token count by one from the current owner's count, and incrementing the token count by one for the receiver accounts. The `to` address will be the new owner of the tokens. After successful execution, this function emits the `Transfer` event.

The _checkOnERC721Received internal function

The `_checkOnERC721Received()` function is an `internal` function that is responsible for further calling the `onERC721Received()` function on an `ERC721Receiver` contract. Here, the `ERC721` contract is behaving like a proxy to the `ERC721Receiver` contract. If the recipient is not a contract account, the function simply returns `true`.

The function takes the following arguments:

- `from`: The address of the current owner of the NFT
- `to`: The address of the recipient of the NFT
- `tokenId`: The `tokenId` of the NFT that is to be transferred

Let's look at the `_checkOnERC721Received()` function definition:

```
function _checkOnERC721Received(address from, address to, uint256 tokenId,
    bytes memory _data)
    internal returns (bool)
{
    if (!to.isContract()) {
        return true;
    }

    bytes4 retval = IERC721Receiver(to).onERC721Received(msg.sender,
        from, tokenId, _data);
    return (retval == _ERC721_RECEIVED);
}
```

As you can see in the preceding code, when the `to` address (the recipient's address) is not a contract address, the function returns `true`. However, if the `to` address is a contract, it further calls the `onERC721Received()` function on the contract. As per the `ERC721Receiver` contract, the call to the `onERC721Received()` function should return the `bytes4` function signature that represents the `onERC721Received()` function signature. The function signature is stored in the `_ERC721_RECEIVED` constant. If a recipient contract does not have `onERC721Received()` function defined, the transaction would fail.

The _clearApproval internal function

`_clearApproval()` is a `private` function that is used by the `ERC721` implementation itself to clear any approvals given previously for the provided `tokenId` variable. The function takes the `tokenId` for which the approvals should be cleared. The function is defined as follows:

```
function _clearApproval(uint256 tokenId) private {
    if (_tokenApprovals[tokenId] != address(0)) {
        _tokenApprovals[tokenId] = address(0);
    }
}
```

As it is a `private` function, it cannot be accessed from the inherited contracts. However, this is only used by the `ERC721` implementation itself. The function first verifies that the current approval address is not `address(0)`; if not, it clears out the approval by setting it to `address(0)`.

Initially, the token approvals for a new `ERC721` token is the default address value, meaning `address(0)`. Once the owner of the token gives approval to another address, the owner calls the `approve()` function. This operation gives another address the token approval of the owner's token. Once the token is burned or transferred to some other address, the approval must be cleared; hence, internally, the `_clearApproval()` function is called, which resets the address back to `address(0)` if it was previously set to a non-zero address.

Events

The ERC721 standard has mainly three events, Transfer, Approval, and ApprovalForAll. These events are emitted from the different function calls of the ERC721 implementation, as we have seen in the ERC721 implementation section. For example, the _mint() function triggers the Transfer event upon successful execution of the function.

The Transfer event is triggered when any token-transfer-related function is called. However, for approvals, there are two types of approval an owner can give:

- **Single Token Approval**: The owner of the token gives approval for a single NFT to another address. This triggers the Approval event.
- **Operator Approval**: The owner can give access to all of the tokens they own to an operator. This triggers the ApprovalForAll event.

As the blockchain is a series of transactions, the events are also stored along with the transaction in which they are triggered. The client application or DApps can listen for new events or filter out from the previously emitted events on the blockchain. And based upon that, it can trigger the corresponding action at the client side or at off-chain level.

Let's look at each of the event definitions. Let's start with the Transfer event.

The Transfer event

The Transfer event is triggered when an NFT is transferred from the current owner's account to the receiver's account.

The event takes three arguments:

- from: The address from where the token is transferred
- to: The address of the receiver of the token
- tokenId: The tokenId of the NFT that is transferred

All of the preceding arguments are indexed in the event, which would allow the client applications to filter out the events:

```
event Transfer(address indexed from, address indexed to, uint256 indexed tokenId);
```

The event is also triggered when the new NFT is minted, or an existing NFT is burned from the contract. When a new NFT is minted, the `from` address will be `address(0)`. However, when an existing NFT is burned, the `to` address will be `address(0)`. The following is a snippet from the corresponding functions.

The `Transfer` event is triggered from the `_mint()` function, where the `from` address is `address(0)`:

```
emit Transfer(address(0), to, tokenId);
```

The `Transfer` event is triggered from the `_burn()` function, where the `to` address is `address(0)`:

```
emit Transfer(owner, address(0), tokenId);
```

The `Transfer` event is triggered from the following functions of the `ERC721` implementation:

- `_mint()`: Emits a `Transfer` event when a new NFT is minted
- `_burn()`: Emits a `Transfer` event when an existing NFT is burned
- `_transferFrom()`: Emits a `Transfer` event when an NFT is transferred via any of the transfer operation-related function calls, such as calls to the `transferFrom()` and `safeTransferFrom()` functions

The Approval event

The `Approval` event is triggered when the approval of a token is given or changed.

The event takes three arguments:

- `owner`: The address of the approver who is the owner of the token
- `approved`: The address that will be approved for the given token
- `tokenId`: The `tokenId` of the NFT for which the approval is initiated

All of the preceding arguments are `indexed` in the event, which would allow the client applications to filter out the events:

```
event Approval(address indexed owner, address indexed approved, uint256 indexed tokenId);
```

The `Approval` event is triggered only from the `approve()` function of the `ERC721` implementation.

The ApprovalForAll event

The ApprovalForAll event is triggered when an owner gives or changes approval for all of their NFTs to an operator.

The event takes three arguments:

- owner: This is the address of the owner who is giving approval to an operator.
- operator: This is the address of the operator who will receive full rights for the owner's tokens.
- approved: This is a Boolean value representing that approval is given or revoked. When the event parameter is true, the operator receives the approval; when it is false, the approval is revoked.

Let's look at the ApprovalForAll event definition code:

```
event ApprovalForAll(address indexed owner, address indexed operator, bool
approved);
```

The ApprovalForAll event is triggered only from the setApprovalForAll() function of the ERC721 implementation.

The ERC721TokenReceiver interface

As the name suggest, the ERC721TokenReceiver interface is to receive the callback for an ERC721 NFT. If you are writing a contract that would receive some NFTs, the ERC721TokenReceiver interface may be inherited in your contract. So, it would enable your contract to get a callback function call from the ERC721 NFT contract to your contract. When someone transfers an ERC721 NFT to your contract, you would receive the callback call to the onERC721Received() function, and you would be able to perform some action accordingly. However, the pre-requisite for this callback is that the ERC721 token sender must call the safeTransferFrom() function present on the ERC721 implementation.

If your receiving contract does not inherit from the ERC721TokenReceiver contract and does not have the onERC721Received() function present in its code, then when an ERC721 token is sent to this contract, it would receive the token. However, the callback function call to the onERC721Received() function will not be received by the receiving contract. To perform this, an ERC721 token sender should call the transferFrom() function of the ERC721 implementation.

Let's look at the `ERC721TokenReceiver` interface code:

```
interface ERC721TokenReceiver {
    function onERC721Received(address _operator, address _from, uint256
        _tokenId, bytes calldata _data) external returns(bytes4);
}
```

The interface has only a single `onERC721Received()` function present, and the function takes the following arguments:

- `_operator`: The address of the operator, which means the account who called the function present on the `ERC721` token contract
- `_from`: The address of the previous owner of the NFT
- `_tokenId`: The `tokenId` of the NFT that is transferred
- `_data`: The data as `bytes`, which contains the further functions to be called by the recipient contract

The OpenZeppelin also provides the interface definition for `ERC721TokenReceiver`. You can find `IERC721Receiver.sol` present in the library. You can also find the interface definition on our GitHub repository for this chapter, which is taken from the OpenZeppelin library itself.

The ERC721Metadata interface

The `ERC721Metadata` interface is an optional interface to add other metadata details to your `ERC721` NFTs. The developers can choose to add a token name, a token symbol, and a token URI as metadata to an `ERC721` token. If they want to use these metadata fields for their token, they can use the `ERC721Metadata` contract; otherwise, they can ignore this. The interface provides the following `view` functions:

- `name()`: The function returns the full name of the `ERC721` NFT as a `string` datatype.
- `symbol()`: The function returns the token symbol of the `ERC721` NFT as a `string` datatype.
- `tokenURI()`: The function returns the **Uniform Resource Identifier (URI)** as a `string` datatype. The URI may contain a location to a JSON file. The JSON schema file is present on GitHub, at `https://github.com/PacktPublishing/Mastering-Solidity/blob/master/Chapter08/json_schema/ERC721MetadataSchema.json`.

Let's look at the ERC721Metadata interface code:

```
interface ERC721Metadata is ERC721 {
    function name() external view returns (string memory _name);
    function symbol() external view returns (string memory _symbol);
    function tokenURI(uint256 _tokenId) external view returns
    (string memory);
}
```

OpenZeppelin also provides the interface definition for ERC721Metadata; you can find IERC721Metadata.sol and ERC721Metadata.sol present in the library. You can also find the interface definition and implementation on our GitHub repository for this chapter, which is taken from the OpenZeppelin library itself.

The ERC721 enumerable

The ERC721Enumerable interface is an optional interface for adding more features to the ERC721 NFT.

The interface defines the following view functions:

- totalSupply(): The function returns the total number of NFTs issued by the contract that are not burned. Also, each token has a valid owner address that is not address(0).
- tokenByIndex(): The function returns the tokenId of an NFT that is stored at the provided index.
- tokenOfOwnerByIndex(): The function returns the tokenId of an NFT that is stored at a particular index of the owner's list of tokens.

Let's look at the ERC721Enumerable interface code:

```
interface ERC721Enumerable is ERC721 {
    function totalSupply() external view returns (uint256);
    function tokenByIndex(uint256 _index) external view returns (uint256);
    function tokenOfOwnerByIndex(address _owner, uint256 _index) external
    view returns (uint256);
}
```

OpenZeppelin also provides the interface definition for ERC721Enumerable; you can find IERC721Enumerable.sol and ERC721Enumerable.sol present in the library. You can also find the interface definition and implementation on our GitHub repository for this chapter, which is taken from the OpenZeppelin library itself.

The ERC721 full implementation

In an ERC721 NFT full implementation, the basic ERC721 functions and events are supported. It also supports the features of the ERC165, ERC721Enumerable, and ERC721Metadata contracts. The full implementation in the OpenZeppelin library is defined as follows in the ERC721Full.sol file:

```
import "./ERC721.sol";
import "./ERC721Enumerable.sol";
import "./ERC721Metadata.sol";

contract ERC721Full is ERC721, ERC721Enumerable, ERC721Metadata {
    constructor (string memory name, string memory symbol)
        public
        ERC721Metadata(name, symbol)
    {
    }
}
```

As shown in the preceding code, ERC721Full inherits from the implementations of ERC721, ERC721Enumerable, and ERC721Metadata.

Summary

In this chapter, we have seen the ERC721 NFT standard, which is a bit different from the ERC20 token standard. Both the ERC20 and ERC721 standards serve different purposes and applications. ERC20 should be used when you need fungible tokens and each token has the same economic value. However, the ERC721 token should be used when you would need each token to be unique, and each of them can have different economic values.

We have seen the ERC721 NFT standard implementation that the OpenZeppelin 2.1.1 library provides. As a developer, you must keep checking the OpenZeppelin library for any recent fixes or improvements to the ERC721 source code. There are many other libraries and contract files that OpenZeppelin provides, which we will have a deeper look into in the next chapter.

Questions

1. What is the use of the ERC165 standard?
2. Where should the ERC721 NFT standard be used?
3. Why are the _mint() and _burn() functions present in the ERC721.sol file?
4. Are there any advanced token standards for non-fungible tokens?
5. Why isn't there a transfer() function in ERC721 like the one in the ERC20 standard?
6. How can we find out whether a new token has been minted, or an existing token has been burned, using the Transfer event?
7. Why is there a bytes data parameter present in the IERC721Receiver.onERC721Received() function?

9
Deep Dive into the OpenZeppelin Library

We have been referring to the OpenZeppelin libraries in some places in Chapter 3, *Control Structures and Contracts*, in Chapter 4, *Learning MetaMask and Remix,* and in the sample Solidity code we discussed in those chapters. In this chapter, we will look into the Solidity contract files that OpenZeppelin provides. The OpenZeppelin Solidity contract files and libraries are open source. These have been tested, audited, and used in many Ethereum **Decentralized Applications (DApps)**. The OpenZeppelin project files are maintained by the Zeppelin company, and the libraries were developed to ensure high-quality code and security standards.

After completing this chapter, you will be able to install and use the OpenZeppelin library contract files in your project. You will be able to decide which library files you want to bring in for use according to the architecture of your project. You will also get some idea of how to write your own Crowdsale contract that accepts ether and mints new tokens.

We will cover the following topics in this chapter:

- Installation and usage of the OpenZeppelin contract and library files
- Using ownership-related contract files to provide controlled access to contracts
- Using role-based contract files to provide role-based access to contracts
- Using ERC20 token standard-related files to create tokens according to your needs
- Using the SafeMath library to safeguard against integer overflow and underflow attacks
- Using Crowdsale contract files to build contracts for token sale or crowdfunding
- Using some utility contract files

Technical requirements

OpenZeppelin has the `openzeppelin-solidity` project, and the source for the project is on GitHub. The GitHub link for this project is `https://github.com/OpenZeppelin/openzeppelin-solidity`.

The OpenZeppelin library should be installed using the `npm` command; it should not be copy-pasted into your contracts folder. However, to discuss some of these files, we have kept them in a `Chapter09` folder that's present on GitHub: `https://github.com/PacktPublishing/Mastering-Blockchain-Programming-with-Solidity/tree/master/Chapter09`.

In this chapter, we are going to use OpenZeppelin v2.1.1 project files. These files are compatible with the following:

- Solidity v0.5.0 or later
- Truffle v5.0.0 or later
- Node.js v8.9.4 or later

The OpenZeppelin project

The OpenZeppelin project is built and maintained by a company called Zeppelin. The project is open source—anyone can use the Solidity contract files that are released by the project. The project contains many contract files that are mostly used by DApps to build upon and maintain the contract architecture. Many of the contract files that are present in the project have been battle-tested and are secure enough to be used in your project.

However, as a developer, you need to keep checking the issues that have been fixed on the OpenZeppelin GitHub project. You should always check the new OpenZeppelin releases and read the bugs list, which have been fixed in the new release. If a bug has been fixed in a contract file that you have used in your project, verify and ensure that your project contract code hasn't been affected; otherwise, you won't know how it could impact the contracts that are present in production.

Once the OpenZeppelin project has been installed in your Truffle project using the `npm` command (follow the installation instructions shown in further sections of this chapter), the library files will be installed in the `node_modules/openzeppelin-solidity/contracts` folder. You will find that the library contract files are placed in different folders according to the contract's behavior.

Let's look into each folder:

- `access` and `access/roles`: These folders contain contract files that are related to role-based access. For example, the `MinterRole.sol` contract should be used to assign `MinterRole` to one or multiple whitelisted addresses. An address that has a `MinterRole` role assigned to it can mint new tokens.

- `crowdsale`: This folder contains subfolders that have contract files that you can use to build different types of `crowdsale` features. For example, `WhitelistedCrowdsale.sol` allows every participant of `crowdsale` to be a whitelisted address.

- `cryptography`: This folder contains contracts that perform cryptographic operations within the contract. For example, a user can sign a message off-chain using the private key of an account. To recover the public key of the signer from this signed message on-chain, the `ECDSA.sol` contract library is used.

- `drafts`: This folder contains different types of contracts that are still in the draft stage. You can use these contracts in your project, but the code has not undergone heavy auditing (it's not battle-tested code), and therefore is unlikely to be safe to use.

- `examples`: This folder contains some example contracts to showcase how the OpenZeppelin library can be used when you want to build your own contracts.

- `introspection`: This folder contains contracts that help in examining the type or property of a contract. For example, currently, this folder contains an ERC165 standard and its implementation, which is used to check that a specific method is present in the called contract. In the future, there could be more such contracts added to the folder.

- `lifecycle`: This folder contains life cycle-related contract files. Currently, it contains a single file, `Pausable.sol`, which is used to pause/unpause some of the functionality of the implementing contract.

- `math`: This folder contains library files that should be used for mathematical arithmetic calculations. For example, using `SafeMath.sol` for arithmetic calculations would prevent integer overflow and underflow attacks.

- `mocks`: The files that are present in this folder are only mock contracts. They are used for testing OpenZeppelin contract files. These mock contract files can be used for testing your contracts locally or on testnet. However, the contract code that you will deploy on mainnet should not be dependent on these mock contract files at the time of deployment or after.

- `ownership`: The contracts present in this folder provide restricted access to the functions of the implementing contract. For example, the `Ownable.sol` contract is used when some functions of the implementing contract should be called by the owner only.
- `payment`: The contracts that are present in this folder are used for carrying out ether payment-related features. For example, using `Escrow.sol` allows you to deposit ether into an escrow contract, which acts as a judge between two or multiple parties and holds funds.
- `token/ERC20`: This folder contains all of the ERC20 token standard-related interfaces and contracts. For example, `ERC20Pausable.sol` can be used when the owner of the contract wants to pause/unpause ERC20 function calls.
- `token/ERC721`: This folder contains all of the ERC721 non-fungible token, standard-related interfaces and contracts. For example, `ERC721Burnable.sol` can be used to allow the token holder to burn their own ERC721 tokens.
- `utils`: This folder contains some utility libraries and contracts. For example, the `Address.sol` library can be used to check whether an address is a contract address or not.

In this chapter, we will discuss a few libraries and contract files of the OpenZeppelin project that are mostly used for decentralized application development.

For your benefit, we have kept the OpenZeppelin library files under the `Chapter09/contracts` folder. We will only discuss these files in this chapter. However, you should not copy-paste files from the `Chapter09/contract` folder into your projects, as this is not the recommended way. Follow the installation steps that are discussed in the following section to use the OpenZeppelin library files.

The OpenZeppelin library has a defined format for state variable and function names. Any variable or function name prefixed with _ (underscore) either has `private` or an `internal` visibility. If there aren't any prefixes, it means they have either `public` or `external` visibility.

Let's learn how to install the OpenZeppelin library in our Truffle project.

 To get all the OpenZeppelin library files released in each version, you can download the `.zip` file for each release from `https://github.com/OpenZeppelin/openzeppelin-solidity/releases`.

Installation

To install OpenZeppelin project files into your Truffle project, you should have your Truffle project directory configured. In the root folder of your Truffle project, run the following command:

```
$ npm init -y
```

The preceding command is optional. If your project folder already has npm initialized and the package.json file is present, you can skip this step.

The following command will install the latest stable version of the OpenZeppelin project and keep all library contract files under the node_modules/openzeppelin-solidity/contracts folder. By using the --save-exact option, the installation ensures that openzeppelin-solidity will pin to that version. If this option isn't used, then your OpenZeppelin files will be upgraded to a more recent version the next time you execute the npm install command:

```
$ npm install --save-exact openzeppelin-solidity
```

Now that we have installed the latest version of the OpenZeppelin library files into our Truffle project, let's learn how to use the library files in our Solidity contracts.

Usage

Once the OpenZeppelin library has been installed in your Truffle project, you will be able to refer to the OpenZeppelin library contracts in your Solidity contract files.

We have seen some of the example contract files in the previous chapter. Some of these contract files already use OpenZeppelin library files, such as the DeploymentExample.sol file we discussed in Chapter 4, *Learning MetaMask and Remix*, and the contract file that's present in the Chapter04/contracts folder:

```
import "openzeppelin-solidity/contracts/ownership/Ownable.sol";
import "openzeppelin-solidity/contracts/math/SafeMath.sol";

contract DeploymentExample is Ownable {
    using SafeMath for uint;

    //...
}
```

As shown in the preceding code, you can import OpenZeppelin library contract files using the `import` statement. The preceding example imports the `Ownable.sol` and `SafeMath.sol` contract files from the OpenZeppelin library and uses them in the `DeploymentExample.sol` contract.

As we mentioned previously, the contract files are present under `<ProjectRoot>/node_modules/openzeppelin-solidity/contracts`; however, in the `import` statement, you don't have to provide the absolute path of the contract file. The Truffle framework takes care of this and allows you to import a contract that's present under the `node_modules` folder. However, there could be other development frameworks that don't have this feature in order to auto-refer the OpenZeppelin files from the `node_modules` folder, such as Embark.

When you are using the Remix IDE in your web browser, you can import the OpenZeppelin library contracts directly by using the GitHub URL. For example, using the following statement in your contract would use the `Ownable` contract from the OpenZeppelin library in your contract:

```
import
"https://github.com/OpenZeppelin/openzeppelin-solidity/contracts/ownership/
Ownable.sol";
```

Now, let's learn about the most commonly used OpenZeppelin contract files.

Contract ownership

There are some contracts that require the ownership of the contract to control some behavior in the contract. These contracts inherit from OpenZeppelin ownership contacts. These types of contracts require ownership rights so that the owner of the contract can access and control some of the features/functions of the contract. Sometimes, the ownership of a contract is defined such that it decides and influences token minting, ether/token flow, and ether/token withdrawals.

The ownership of the contracts should be given to the trusted entity. The following responsibilities need to be taken care of:

- If ownership is given to an EOA, the owner needs to keep the private key of that account secure, in a safe place. Losing the private key would incur the loss of ether/tokens or malicious activity. The owner could lose their private key by forgetting where they kept it, or someone who is unauthorized could steal their private key and perform a malicious activity by using it.

- If the ownership of `OwnedContract` is given to a contract—let's say, `OwnerContract`—the developer must ensure that `OwnerContract` can control the required access rights of `OwnedContract`. This is a safer approach as there is no private key associated with the contracts. However, this should only be used when it is required in your contract architecture. For example, let's say you have deployed an ERC20 token contract called `TKNToken`, which can mint new tokens from its owner. However, the `TKNToken` contract's ownership is given to a `Crowdsale` contract. The `Crowdsale` contract can receive ether and, according to the rate, it calls the `mint()` function on the `TKNToken` contract to mint new tokens for the ether sender. In this example, ownership of `TKNToken` is handled by the `Crowdsale` contract.
- There is another way to manage the ownership of the contracts, and that is by using **multiple signature (multisig)** contracts. By using these contracts, multiple parties control the ownership of another contract. To perform any operation, some of the owners have to sign the transactions. Refer `Chapter 10`, *Using Multisig Wallets*, for details on how to control a contract using multisig contract.

Let's start by learning about the `Ownable.sol` contract.

Contract ownership using Ownable.sol

The `Ownable.sol` contract provides the most basic single account ownership to a contract. Only one account will become the owner of the contract and can perform administration-related tasks. The current owner of the contract can either transfer or renounce the ownership of the contract. This type of access control should be used wherever single ownership is required.

The OpenZeppelin contract present at `Chapter09/openzeppelin-solidity/contracts/ownership/Ownable.sol` is defined as follows:

```
contract Ownable {
    address private _owner;

    event OwnershipTransferred(address indexed previousOwner, address
    indexed newOwner);

    constructor () internal {
        _owner = msg.sender;
        emit OwnershipTransferred(address(0), _owner);
    }

    function owner() public view returns (address) {
```

```
        return _owner;
    }

    modifier onlyOwner() {
        require(isOwner());
        _;
    }

    function isOwner() public view returns (bool) {
        return msg.sender == _owner;
    }

    function renounceOwnership() public onlyOwner {
        emit OwnershipTransferred(_owner, address(0));
        _owner = address(0);
    }

    function transferOwnership(address newOwner) public onlyOwner {
        _transferOwnership(newOwner);
    }

    function _transferOwnership(address newOwner) internal {
        require(newOwner != address(0));
        emit OwnershipTransferred(_owner, newOwner);
        _owner = newOwner;
    }
}
```

Let's review the preceding code in more detail. This contract contains the following information:

- **State variables:**
  - _owner: This variable stores the current owner's address of the contract.
- **Events:**
  - OwnershipTransferred: This event is fired essentially three times. When ownership of the contract is assigned at the time of deployment, the current owner transfers the ownership to another address or renounces it.

- **Modifiers:**
  - onlyOwner: The implementing contract can use this function modifier to allow only owner to call that defined function.
- **Constructor:**
  - In the constructor, the contract assigns ownership to the deployer of the contract (msg.sender). Hence, the deployer becomes the owner of the contract.

- **Functions:**
  - owner(): This `public view` function returns the current owner's address.
  - isOwner(): This `public view` function checks whether the caller of this function has ownership assigned and returns `true`/`false` in response.
  - renounceOwnership(): This is a `public` function and, by using this, only the current owner is allowed to leave their ownership. If the current owner calls this function, they cannot claim ownership of the contract later on. This means that the contract won't have an owner assigned who can control `onlyOwner` functions. This puts the contract in a dangerous situation when the ownership of the contract is lost unintentionally.
  - transferOwnership(): This is a `public` function; the current owner can transfer ownership to another address by calling this function. By doing this, the current owner gives up their ownership of the contract.
  - _transferOwnership(): This is an `internal` function that's used to transfer ownership. The implementing contract can also use this function in special cases when they need to define custom controls.

Claimable ownership using Claimable.sol

In this chapter, our major focus is on OpenZeppelin version 2.1.1 contract files. However, in the older version of OpenZeppelin, 1.12.0, `Claimable.sol` was also present as a contract. This contract can be useful for developers, and so we discuss this contract in this chapter. In the new version, 2.0.0, it has been removed. One good feature of the `Claimable.sol` contract is that it supports transferring ownership in a two-step process:

1. The current owner calls the `transferOwnership()` function to give ownership to another address, let's say, `0xNewOwnerAddress`. The ownership isn't transferred; rather, it is in a pending state.
2. The new owner address, `0xNewOwnerAddress`, has to call the `claimOwnership()` function to gain ownership of the contract.

The OpenZeppelin contract present at `Chapter09/openzeppelin-solidity/contracts/ownership/Claimable.sol` is defined as follows:

```
contract Claimable is Ownable {
  address public pendingOwner;

  modifier onlyPendingOwner() {
    require(msg.sender == pendingOwner);
    _;
```

```
  }

  function transferOwnership(address newOwner) public onlyOwner {
    pendingOwner = newOwner;
  }

  function claimOwnership() public onlyPendingOwner {
    _transferOwnership(pendingOwner);
    pendingOwner = address(0);
  }
}
```

Let's review the preceding code in more detail. This contract contains the following functions and modifiers:

- **Inherits from:**
  - Ownable: This inherits modifiers and internal functions from Ownable.sol since those are being used in the contract.
- **Modifier:**
  - onlyPendingOwner: This is the modifier that allows only the pending owner to call the function. For example, this is used in the claimOwnership() function, which only the pending owner can call.
- **Functions:**
  - transferOwnership(): The current owner is allowed to call this function to initiate the transfer of ownership to a new address. However, the ownership is not transferred; it will be in a pending state.
  - claimOwnership(): If the pendingOwner variable is set with a valid address, that address initiates the call to this function to claim their ownership. After a successful transaction, the old owner will be removed from ownership and the pendingOwner address will become the new owner of the contract. This function also resets the pendingOwner state variable by setting it to address(0).

Roles library

The Roles library is used to store and manage different kinds of role-based access for contracts. As per your contract architecture needs, you can assign different kinds of roles to the different contracts or individual EOA accounts.

All of the role-related files are in the `Chapter09/openzeppelin-solidity/contracts/access` folder of the OpenZeppelin library.

Let's start by learning about a basic roles contract called `Roles.sol`.

Managing roles using Roles.sol

`Roles.sol` is a generic library that stores the access rights that are provided to an address. All the roles-based contracts defined in OpenZeppelin use this library. It also exposes some `internal` functions that can only be called within the contract. Using this generic library, we can create and define our own roles for different kinds of access controls. In this section, we will look into the different roles that OpenZeppelin provides, such as `PauserRole.sol` and `MinterRole.sol`, which also inherit from `Roles.sol`.

The OpenZeppelin contract present at `Chapter09/openzeppelin-solidity/contracts/access/Roles.sol` is defined as follows:

```solidity
library Roles {
    struct Role {
        mapping (address => bool) bearer;
    }

    function add(Role storage role, address account) internal {
        require(account != address(0));
        require(!has(role, account));

        role.bearer[account] = true;
    }

    function remove(Role storage role, address account) internal {
        require(account != address(0));
        require(has(role, account));

        role.bearer[account] = false;
    }

    function has(Role storage role, address account) internal view returns (bool) {
        require(account != address(0));
        return role.bearer[account];
    }
}
```

Let's review the preceding code in more detail. The library contract contains the following functions and structs:

- **Structs:**
  - The `Role` struct is defined in the library. It has a mapping to store the addresses that have the role enabled or disabled. If the Boolean value is set to `true` for the respective address, the role is enabled, otherwise disabled.
- **Functions:** All of these functions are `internal` functions that will be called from the contracts using this library:
  - `add()`: The function to assign the role to a new address
  - `remove()`: The function to revoke the assigned role from an existing address
  - `has()`: The `view` function to check whether an address has the role

There are many predefined roles present in the OpenZeppelin library. These roles are also used in other contracts for role-based access. Let's look at these now.

Manage PauserRole using PauserRole.sol

The `PauserRole.sol` contract is used to manage `PauserRole` for a contract or an EOA. An address with `PauserRole` assigned to it can pause/unpause contract features and functions. In the upcoming section, we will look into the `Pausable.sol` contract and learn how to use this role for this purpose.

The OpenZeppelin contract present at `Chapter09/openzeppelin-solidity/contracts/access/roles/PauserRole.sol` is defined as follows:

```
contract PauserRole {
    using Roles for Roles.Role;

    event PauserAdded(address indexed account);
    event PauserRemoved(address indexed account);

    Roles.Role private _pausers;

    constructor () internal {
        _addPauser(msg.sender);
    }

    modifier onlyPauser() {
        require(isPauser(msg.sender));
        _;
```

```
    }

    function isPauser(address account) public view returns (bool) {
        return _pausers.has(account);
    }

    function addPauser(address account) public onlyPauser {
        _addPauser(account);
    }

    function renouncePauser() public {
        _removePauser(msg.sender);
    }

    function _addPauser(address account) internal {
        _pausers.add(account);
        emit PauserAdded(account);
    }

    function _removePauser(address account) internal {
        _pausers.remove(account);
        emit PauserRemoved(account);
    }
}
```

As you can see, when the PauserRole contract is deployed, a constructor is called and assigns the contract deployer with the PauserRole role. Now, the same contract deployer address is allowed to assign more PauserRole roles to any new addresses. Hence, there can be many addresses with PauserRole. However, if anyone wants to leave PauserRole, they can call the renouncePauser() function to revoke their access.

The contract contains the following functions, events, modifier, constructor, and state variables:

- **State variables**:
    - _pausers: This is a type of Roles.Role struct, which we already discussed in the *Roles* section. All the functions of the Roles library apply to this variable.
- **Events**:
    - PauserAdded: This event will be triggered when a new PauserRole is assigned.
    - PauserRemoved: This event will be triggered when an existing PauserRole is revoked.

- **Constructor**:
  - In the constructor, the contract assigns `PauserRole` to the deployer of the contract.
- **Modifier**:
  - `onlyPauser`: This modifier only allows a function to be executed if `msg.sender` has `PauserRole` assigned.
- **Functions**:
  - `isPauser()`: This `public view` function checks whether a given address has `PauserRole` or not.
  - `addPauser()`: This is a `public` function assigns a new address with `PauserRole`. This can only be called by an address that has `PauserRole` already.
  - `renouncePauser()`: This is a `public` function that allows anyone that has the `PauserRole` assigned call the function to revoke themselves from `PauserRole`. It only removes `msg.sender` from `PauserRole`, and so they don't have the rights to remove another address that has `PauserRole`.
  - `_addPauser()`: This is an `internal` function that assigns an address with `PauserRole`.
  - `_removePauser()`: This is an internal function that revokes `PauserRole` access from an address.

Other roles

There are some other role-specific contracts present in the OpenZeppelin library that are defined just like `PauserRole.sol`, which we discussed previously. The functions, modifiers, and events are almost similar for each role—only the usage is different, and so you can use these functions in development very easily. Let's discuss the purpose of these files in more detail:

- `MinterRole.sol`: This is the contract used to manage `MinterRole`. An address that's assigned with `MinterRole` is allowed to mint new tokens. This role can be used in pre-crowdsale to mint new tokens for pre-**Initial Coin Offering (ICO)** investors.
- `WhitelistAdminRole.sol`: This is the contract used to manage `WhitelistAdminRole`. An address that's assigned with `WhitelistAdminRole` and is allowed to whitelist any address. This is used in whitelisted crowdsales.

- `WhitelistedRole.sol`: This is the contract used to manage `WhitelistedRole`. An address assigned `WhitelistedRole` is allowed to perform certain operations on a contract. This is mostly used to allow only whitelisted addresses to invest in a crowdsale.

 Pre-ICO is a phase in which a company or organization goes to private investors to raise funding (pre-crowdsale) before going to public ICO. In most cases, in the pre-ICO phase, a company raises funds off-chain and mints tokens for their investors on-chain via mintable contracts.

Life cycle contracts

The `lifecycle` contract controls the life cycle flow of functions. For example, using the `Pausable` contract, we can do the following:

- Pause/unpause the minting of new tokens
- Pause/unpause the transfer of tokens
- Pause/unpause the acceptance of ether

In OpenZeppelin 2.1.1, only one `Pausable.sol` contract is present in the `Chapter09/openzeppelin-solidity/contracts/lifecycle` folder. In the future, more contracts may be added.

Pause/unpause functions using Pausable.sol

As we discussed previously, there are some examples of the `Pausable` contract where it can be used to control the life cycle of a contract. We can use `Pausable` where administrative functionality is needed.

However, if you use the `Pausable` contract, then the `PauserRole` role has to be assigned to the account(s) that you can trust to pause or unpause the contract. If you are using the `Pausable` contract in your decentralized application, then it might not be called a pure decentralized application since it has a centralized feature that can pause/unpause functionality of contract. However, according to your system architecture, you can use or avoid using the `Pausable` contract.

The OpenZeppelin contract present at `Chapter09/openzeppelin-solidity/contracts/lifecycle/Pausable.sol` is defined as follows:

```
contract Pausable is PauserRole {
    event Paused(address account);
    event Unpaused(address account);

    bool private _paused;

    constructor () internal {
        _paused = false;
    }

    function paused() public view returns (bool) {
        return _paused;
    }

    modifier whenNotPaused() {
        require(!_paused);
        _;
    }

    modifier whenPaused() {
        require(_paused);
        _;
    }

    function pause() public onlyPauser whenNotPaused {
        _paused = true;
        emit Paused(msg.sender);
    }

    function unpause() public onlyPauser whenPaused {
        _paused = false;
        emit Unpaused(msg.sender);
    }
}
```

Let's review the preceding code in more detail. The contract contains the following functions, events, modifiers, and constructor and state variables:

- **Inherits from**:
  - `PauserRole`: The `Pausable` contract inherits from `PauserRole` in order to allow a `PauserRole` to pause/unpause a function.
- **State variables**:
  - `_paused`: A Boolean state variable to store the paused state.

- **Events**:
  - `Paused`: An event is triggered when the contract is paused by an address that has `PauserRole`.
  - `Unpaused`: An event is triggered when the contract is unpaused by an address that has `PauserRole`.
- **Modifiers**:
  - `whenNotPaused`: The function modifier ensures that the contract is not in the paused state, then only allows function execution.
  - `whenPaused`: The function modifier ensures that the contract is in the paused state, then only allows function execution.
- **Constructor**:
  - The contract is created and sets the default unpaused state by setting the _paused state variable to `false`.
- **Functions**:
  - `paused()`: This `public view` function returns a Boolean representing the state of the pause/unpause. When the function returns `true`, it means it's paused; otherwise, it isn't.
  - `pause()`: This `public` function puts the contract in paused state. This can only be called by an address that has `PauserRole`, and only if the contract is not in the paused state.
  - `unpause()`: This `public` function puts the contract in unpaused state. This can only be called by an address that has `PauserRole`, and only if the contract is in the paused state.

The ERC20 token standard library

ERC20 is a token standard that's used in the Ethereum blockchain. It defines some standard functions and events that should be implemented by a contract if there is a need for a token.

The OpenZeppelin library includes contract files for the full implementation of the ERC20 token standard. Some other relevant ERC20 feature-related files are also present; using these files, you can decorate your ERC20 token contract and add more optional features to it. For example, with the `Mintable` and `Pausable` features, you can add token-minting and token-transfer-pause features to your ERC20 token contract.

All of the roles-related files are present in the `Chapter09/openzeppelin-solidity/contracts/token/ERC20` folder of the OpenZeppelin library.

Let's start by learning about the ERC20 interface.

ERC20 interface – IERC20.sol

The OpenZeppelin library includes the ERC20 interface file, `IERC20.sol`. The `IERC20.sol` file is an interface that defines the functions and events that are required for the ERC20 token standard. We discussed the ERC20 token standard interface in detail in `Chapter 7`, *ERC20 Token Standard*.

The OpenZeppelin interface present at `Chapter09/openzeppelin-solidity/contracts/token/ERC20/IERC20.sol` is defined as follows:

```
interface IERC20 {
    function transfer(address to, uint256 value) external returns
    (bool);

    function approve(address spender, uint256 value) external returns
    (bool);

    function transferFrom(address from, address to, uint256 value) external
    returns (bool);

    function totalSupply() external view returns (uint256);

    function balanceOf(address who) external view returns (uint256);

    function allowance(address owner, address spender) external view
    returns (uint256);

    event Transfer(address indexed from, address indexed to, uint256
    value);

    event Approval(address indexed owner, address indexed spender, uint256
    value);
}
```

The contract that's defining the ERC20 token features has to implement `IERC20.sol`, which will enable this inherited contract to define the behavior of the previously listed functions.

Full ERC20 implementation using ERC20.sol

We discussed the ERC20 token standard and its functions in detail in Chapter 7, *ERC20 Token Standard*. You can refer to that chapter for a detailed explanation. In this section, we will look into the OpenZeppelin implementation files for the ERC20 token standard.

The OpenZeppelin contract can be found at Chapter09/openzeppelin-solidity/contracts/token/ERC20/ERC20.sol. You can also refer to the full code of ERC20.sol file that's available on GitHub at https://github.com/PacktPublishing/Mastering-Solidity/blob/master/Chapter09/openzeppelin-solidity/contracts/token/ERC20/ERC20.sol.

The following code is a shortened version:

```
contract ERC20 is IERC20 {
    using SafeMath for uint256;
    mapping (address => uint256) private _balances;
    mapping (address => mapping (address => uint256)) private _allowed;
    uint256 private _totalSupply;

    function totalSupply() public view returns (uint256) { ... }

    function balanceOf(address owner) public view returns (uint256) { ... }

    function allowance(address owner, address spender) public view returns
    (uint256) { ...}

    function transfer(address to, uint256 value) public returns (bool)
    { ... }

    function approve(address spender, uint256 value) public
    returns (bool) {
        require(spender != address(0));

        _allowed[msg.sender][spender] = value;
        emit Approval(msg.sender, spender, value);
        return true;
    }

    function transferFrom(address from, address to, uint256 value) public
    returns (bool) {
        _allowed[from][msg.sender] = _allowed[from][msg.sender].sub(value);
        _transfer(from, to, value);
        emit Approval(from, msg.sender, _allowed[from][msg.sender]);
        return true;
    }
```

```
function increaseAllowance(address spender, uint256 addedValue) public
returns (bool) {
    require(spender != address(0));

    _allowed[msg.sender][spender] =
      _allowed[msg.sender][spender].add(addedValue);
    emit Approval(msg.sender, spender,
    _allowed[msg.sender][spender]);
    return true;
}

function decreaseAllowance(address spender, uint256 subtractedValue)
public returns (bool) {
    require(spender != address(0));

    _allowed[msg.sender][spender] =
      _allowed[msg.sender][spender].sub(subtractedValue);
    emit Approval(msg.sender, spender, _allowed[msg.sender][spender]);
    return true;
}

//Rest of the code
}
```

Let's review the preceding code in more detail. The contract contains the following functions and state variables:

- **Inherits from**:
    - IERC20: This implements the ERC20 interface and provides the definition for each function. This also inherits events declared in IERC20.sol.
- **State variables**:
    - _balances: A Solidity mapping to store the token balance that corresponds to each address that holds the token.
    - _allowed: A Solidity mapping, that keeps track of the approved number of tokens by the approver to the spender.
    - _totalSupply: The variable that stores the total supply of the tokens.
- **Functions**:
    - totalSupply(): A public view function to return the total supply of the tokens.
    - balanceOf(): A public view function to check the token balance of the given address.
    - allowance(): A public view function to check the allowance balance of the token of a spender who has been approved by an approver.

- `transfer()`: A `public` function to transfer the given number of tokens from the caller's balance to another address.
- `approve()`: A `public` function that approves the number of tokens to a given address of the spender. Later, the spender can spend tokens on behalf of the approver using the `transferFrom()` function.
- `transferFrom()`: A `public` function that is called by a spender and transfers a given number of tokens from an approver's balance to a target address.
- `increaseAllowance()`: A `public` function for the approver to increase the previously allowed token balance of a spender.
- `decreaseAllowance()`: A `public` function for the approver to decrease the previously allowed token balance of a spender.
- `_transfer()`: An `internal` function to transfer a given number of tokens from one address to another.
- `_mint()`: An `internal` function to mint a given number of tokens for an address.
- `_burn()`: An `internal` function to burn a given number of tokens from an address' token balance.
- `_burnFrom()`: An `internal` function to burn a given number of tokens from an approver's token balance, who has previously approved allowances to the spender (the spender is `msg.sender`).

Perform safe ERC20 function calls using SafeERC20.sol

As we have seen, the ERC20 standard token implementation functions also return the transaction status as a Boolean. It's good practice to check for the return status of the function call to ensure that the transaction was successful. It is the developer's responsibility to enclose these function calls with `require()` to ensure that, when the intended ERC20 function call returns `false`, the caller transaction also fails. However, it is mostly missed by developers when they carry out checks; in effect, the transaction would always succeed, even if the token transfer didn't. The following is an example of enclosing a token `transfer` call and an `approve` call within the `require()` function:

```
//When calling transfer
require(ERC20(tokenAddress).transfer(to, value));

//When calling transferFrom
```

```solidity
require(ERC20(tokenAddress).transferFrom(from, to, value));

//When calling approve
require(ERC20(tokenAddress).approve(spender, value));
```

The preceding code shows the `transfer()`, `transferFrom()`, and `approve()` function calls enclosed in `require()` to ensure the contract's safety.

However, OpenZeppelin provides the `SafeERC20.sol` contract to ensure the safety of these calls; it is helpful to protect the contract from unintended behavior.

The OpenZeppelin contract present at `Chapter09/openzeppelin-solidity/contracts/token/ERC20/SafeERC20.sol` is defined as follows:

```solidity
import "./IERC20.sol";
import "../../math/SafeMath.sol";

library SafeERC20 {
    using SafeMath for uint256;

    function safeTransfer(IERC20 token, address to, uint256 value)
    internal {
        require(token.transfer(to, value));
    }

    function safeTransferFrom(IERC20 token, address from, address to,
    uint256 value) internal {
        require(token.transferFrom(from, to, value));
    }

    function safeApprove(IERC20 token, address spender, uint256 value)
    internal {
        // safeApprove should only be called when setting an initial
          allowance,
        // or when resetting it to zero. To increase and decrease it, use
        // 'safeIncreaseAllowance' and 'safeDecreaseAllowance'
        require((value == 0) || (token.allowance(msg.sender, spender) ==
        0));
        require(token.approve(spender, value));
    }

    function safeIncreaseAllowance(IERC20 token, address spender, uint256
    value) internal {
        uint256 newAllowance = token.allowance(address(this),
        spender).add(value);
        require(token.approve(spender, newAllowance));
    }
```

```
function safeDecreaseAllowance(IERC20 token, address spender, uint256
value) internal {
    uint256 newAllowance = token.allowance(address(this),
    spender).sub(value);
    require(token.approve(spender, newAllowance));
}
}
```

Let's review the preceding code in more detail. The contract contains the following library functions:

- safeTransfer(): This calls the transfer() function on the given ERC20 token contract address and ensures that the transfer of tokens is successful by enclosing the call in require().
- safeTransferFrom(): This calls the transferFrom() function on the given ERC20 token contract address and ensures that the transfer of tokens is successful by enclosing the call in require().
- safeApprove(): The approve() method of ERC20 is prone to front-running attacks. Therefore, this method ensures that, when the current allowance is non-zero, it must be set to zero first before allowing a new allowance.
- safeIncreaseAllowance(): This ensures that the allowance balance is increased safely by enclosing the approve() function with require().
- safeDecreaseAllowance(): This ensures that the allowance balance is decreased safely by enclosing the approve() function with require().

Create tokens with metadata using DetailedERC20.sol

Using the DetailedERC20.sol contract, you can define the optional fields for an ERC20 token; that is, the name, symbol, and decimals of the token.

The OpenZeppelin contract present at Chapter09/openzeppelin-solidity/contracts/token/ERC20/ERC20Detailed.sol is defined as follows:

```
import "./IERC20.sol";

contract ERC20Detailed is IERC20 {

    string private _name;
    string private _symbol;
    uint8 private _decimals;
```

```
constructor (string memory name, string memory symbol, uint8 decimals)
public {
    _name = name;
    _symbol = symbol;
    _decimals = decimals;
}

function name() public view returns (string memory) {
    return _name;
}

function symbol() public view returns (string memory) {
    return _symbol;
}

function decimals() public view returns (uint8) {
    return _decimals;
}
}
```

Let's review the preceding code in more detail. The contract contains the following functions, constructor, and state variables:

- **Inherits from**:
    - IERC20: It inherits ERC20 interface methods
- **State variables**:
    - _name: Stores the name of the token as a string
    - _symbol: Stores the symbol of the token as a string
    - _decimals: Stores the number of decimal places for the token as uint8
- **Constructor**:
    - The constructor takes three arguments—name, symbol, and decimals—and sets them to the corresponding state variables.
- **Functions**:
    - name(): A public view function to return the token's full name
    - symbol(): A public view function to return the token's symbol
    - decimals(): A public view function to return the token's number of decimals

Create mintable tokens using ERC20Mintable.sol

When you have a token that needs to have a mintable feature—so that you are allowed to mint new tokens according to your needs—you can use the ERC20Mintable.sol contract. This contract requires MinterRole which is given to the deployer of the contract. An account that has MinterRole can call the mint() function and mints new tokens to the specified address.

The OpenZeppelin contract present at Chapter09/openzeppelin-solidity/contracts/token/ERC20/ERC20Mintable.sol is defined as follows:

```
import "./ERC20.sol";
import "../../access/roles/MinterRole.sol";

contract ERC20Mintable is ERC20, MinterRole {

    function mint(address to, uint256 value) public onlyMinter returns
    (bool) {
        _mint(to, value);
        return true;
    }
}
```

Let's review the preceding code in more detail. The contract contains the following functions:

- **Inherits from**:
  - ERC20: Inherits the ERC20 token standard function; this also contains the definition of the _mint() function
  - MinterRole: Inherits MinterRole access rights that can be used by the accounts that have MinterRole assigned
- **Functions**:
  - mint(): Mints a given number of tokens to the specified address

Allow token burning using ERC20Burnable.sol

When you need to allow users to burn their tokens, you should use the ERC20Burnable.sol contract. This contract provides two functions: burn() and burnFrom().

The OpenZeppelin contract present at `Chapter09/openzeppelin-solidity/contracts/token/ERC20/ERC20Burnable.sol` is defined as follows:

```
import "./ERC20.sol";

contract ERC20Burnable is ERC20 {

    function burn(uint256 value) public {
        _burn(msg.sender, value);
    }

    function burnFrom(address from, uint256 value) public {
        _burnFrom(from, value);
    }
}
```

Let's review the preceding code in more detail. The contract contains the following functions:

- **Inherits from**:
  - `ERC20`: Inherits the ERC20 token standard function.
- **Functions**:
  - `burn()`: This burns the given number of tokens from the `msg.sender` balance.
  - `burnFrom()`: This burns the given number of tokens from the approver's balance. The caller of this function must have tokens approved from the address given in the function call.

Create pausable tokens using ERC20Pausable.sol

When your token needs to have controlled transfer of tokens, you can use the `ERC20Pausable.sol` contract. This gives the owner of the contract the right to pause or unpause token transfers and approvals. When token transfers are paused, anyone holding the token cannot make `transfer()`, `transferFrom()`, `approve()`, `increaseApproval()`, or `decreaseApproval()` function calls, and calls to these functions would fail. However, when the token is in the unpaused state, these function calls are allowed.

The OpenZeppelin contract present at `Chapter09/openzeppelin-solidity/contracts/token/ERC20/ERC20Pausable.sol` is defined as follows:

```
import "./ERC20.sol";
import "../../lifecycle/Pausable.sol";

contract ERC20Pausable is ERC20, Pausable {

    function transfer(address to, uint256 value) public
    whenNotPaused returns (bool) {
        return super.transfer(to, value);
    }

    function transferFrom(address from, address to, uint256 value) public
    whenNotPaused returns (bool) {
        return super.transferFrom(from, to, value);
    }

    function approve(address spender, uint256 value) public
    whenNotPaused returns (bool) {
        return super.approve(spender, value);
    }

    function increaseAllowance(address spender, uint addedValue) public
    whenNotPaused returns (bool success) {
        return super.increaseAllowance(spender, addedValue);
    }

    function decreaseAllowance(address spender, uint subtractedValue)
    public whenNotPaused returns (bool success) {
        return super.decreaseAllowance(spender, subtractedValue);
    }
}
```

Let's review the preceding code in more detail. The contract contains the following functions:

- **Inherits from**:
  - `ERC20`: Inherits the ERC20 token standard function
  - `Pausable`: Inherits `PauserRole` features and the `whenPaused` and `whenNotPaused` modifiers for use in the contract

- **Functions**:
  - transfer(): Allows token transfer using the ERC20.transfer() function when the contract isn't paused
  - transferFrom(): Allows token transfer using the ERC20.transferFrom() function when the contract isn't paused
  - approve(): Allows the approval of tokens using the ERC20.approve() function when the contract isn't paused
  - increaseAllowance(): Allows you to increase the approval token balance using the ERC20.increaseAllowance() function when the contract isn't paused
  - decreaseAllowance(): Allows you to decrease the approval token balance using the ERC20.decreaseAllowance() function when the contract isn't paused

Math-related libraries

OpenZeppelin provides mathematical calculation-related library files. Using these libraries in your contract greatly improves security and safety against integer overflow and underflow attacks. Integer overflow is when a number is very close to its upper bound and further addition on the number would circle it back to its lower bound value; the opposite is the case with integer underflow. We will discuss integer overflow and underflow in detail in Chapter 14, *Tips, Tricks, and Security Best Practices*, in the *Integer overflow and underflow attack* section.

These libraries have been used in many publicly deployed contracts on the Ethereum blockchain. *These are battle-tested libraries and must be used for any mathematical calculations that are present in the contract.*

The contract library files are in the Chapter09/openzeppelin-solidity/contracts/math folder.

Aggregation functions using Math.sol

This library provides some basic aggregate functions that can be used in a contract.

The OpenZeppelin contract present at `Chapter09/openzeppelin-solidity/contracts/math/Math.sol` is defined as follows:

```
library Math {
    function max(uint256 a, uint256 b) internal pure returns (uint256) {
        return a >= b ? a : b;
    }

    function min(uint256 a, uint256 b) internal pure returns (uint256) {
        return a < b ? a : b;
    }

    function average(uint256 a, uint256 b) internal pure returns (uint256)
    {
        // (a + b) / 2 can overflow, so we distribute
        return (a / 2) + (b / 2) + ((a % 2 + b % 2) / 2);
    }
}
```

Let's review the preceding code in more detail. The contract contains the following library functions:

- `max()`: Returns the maximum value from the given two arguments
- `min()`: Returns the minimum value from the given two arguments
- `average()`: Returns the average value of given two arguments

Arithmetic calculation using SafeMath.sol

The `SafeMath.sol` library provides basic addition, subtraction, multiplication, division, and modulo mathematical calculation functions that can be used in your contract.

The OpenZeppelin contract present at `Chapter09/openzeppelin-solidity/contracts/math/SafeMath.sol` is defined as follows:

```
library SafeMath {
    function mul(uint256 a, uint256 b) internal pure returns (uint256) {
        if (a == 0) {
            return 0;
        }

        uint256 c = a * b;
        require(c / a == b);

        return c;
    }
```

```
function div(uint256 a, uint256 b) internal pure returns (uint256) {
    require(b > 0);
    uint256 c = a / b;
    return c;
}

function sub(uint256 a, uint256 b) internal pure returns (uint256) {
    require(b <= a);
    uint256 c = a - b;

    return c;
}

function add(uint256 a, uint256 b) internal pure returns (uint256) {
    uint256 c = a + b;
    require(c >= a);

    return c;
}

function mod(uint256 a, uint256 b) internal pure returns (uint256) {
    require(b != 0);
    return a % b;
}
}
```

Let's review the preceding code in more detail. This contract contains the following library functions:

- mul(): Multiplies two given uint256 numbers and ensures that there is no integer overflow, otherwise fail transaction.
- div(): Divides the numerator by the denominator. Both are of the uint256 type. This also ensures that the divide by zero error is thrown using the require(b > 0) check. However, by default, EVM throws an error and fails the transaction if the divide by zero error is encountered.
- sub(): Subtracts the given second argument from the first argument. Both are uint256 arguments. This function ensures that there is no integer underflow.
- add(): Adds the two given uint256 numbers and ensures that there is no integer overflow, otherwise fail transaction.
- mod(): Performs a modulo operation on the given uint256 numbers.

Crowdsale

By using crowdsale, we can receive funding from anyone who has ether. A crowdsale is a way of getting ether into the contract and, in return, minting new tokens for these investors. These newly minted tokens may or may not be like the shares of traditional stock markets.

However, this is the part of token economics in which there are some different classifications of tokens, such as security tokens and utility tokens:

- **Security tokens**: These types of tokens are sort of like traditional shares of the stock market. For example, the holders of these tokens get the dividend or profit share of the company that raised the funds and minted tokens for investors. These tokens may or may not support burning the tokens; however, they are linked to the company's performance and their revenue growth. As the company gets an increase in profit, it is shared between the token holders.
- **Utility tokens**: These types of tokens are activity-based tokens. These tokens are consumed while performing a certain activity on the platform that supports these tokens. Normally, the company that raised the funds builds some process or product in which these tokens are consumed and might be burned in a small quantity. By burning these tokens, it reduces the number of tokens that are present in the market and increases the economic value of the tokens in the market.

Later, the newly minted tokens are supported by the centralized and **Decentralized Exchange** (**DEX**) to make them tradable in the market.

The OpenZeppelin library provides many contracts for crowdsale. According to your needs, you can add multiple features to your crowdsale contract.

The contract library files are present under the `Chapter09/openzeppelin-solidity/contracts/crowdsale` folder.

Let's learn about the basic `Crowdsale` contract.

Create crowdsale using Crowdsale.sol

`Crowdsale.sol` is just the basic version of the crowdsale contract. Using this, we can receive ether and mint new tokens for the ether sender. This contract supports the following features:

- Allows the deployer of the contract to set the token rate, wallet address to receive ether, and the pre-minted token contract address.
- Receives ether and send tokens to the sender.
- Provides pre- and post-validation functions for the contracts that inherit from this. The inherited contract can override the definition of these functions in order to pre- and post-validate a crowdsale transaction.

The OpenZeppelin contract is present at `Chapter09/openzeppelin-solidity/contracts/crowdsale/Crowdsale.sol`. Here, we have shortened the code, but you can get the full contract definition on GitHub at `https://github.com/PacktPublishing/Mastering-Blockchain-Programming-with-Solidity/blob/master/Chapter09/openzeppelin-solidity/contracts/crowdsale/Crowdsale.sol`:

```
contract Crowdsale is ReentrancyGuard {

    //State Variables are declared here

    constructor (uint256 rate, address payable wallet, IERC20 token) public
    { ... }

    function () external payable {
        buyTokens(msg.sender);
    }

    function buyTokens(address beneficiary) public nonReentrant payable {
        uint256 weiAmount = msg.value;
        _preValidatePurchase(beneficiary, weiAmount);

        // calculate token amount to be created
        uint256 tokens = _getTokenAmount(weiAmount);

        // update state
        _weiRaised = _weiRaised.add(weiAmount);

        _processPurchase(beneficiary, tokens);
        emit TokensPurchased(msg.sender, beneficiary, weiAmount, tokens);

        _updatePurchasingState(beneficiary, weiAmount);
```

```
        _forwardFunds();
        _postValidatePurchase(beneficiary, weiAmount);
    }

    //Rest of the code here
}
```

Let's review the preceding code in more detail. This contract contains the following state variables, constructor, events, and functions:

- **Inherits from**:
    - ReentrancyGuard: Using the nonReentrant modifier in the buyTokens() function to prevent reentrancy attacks.
- **State variables**:
    - _token: The ERC20 token that is to be minted and sent to the investors
    - _wallet: The wallet address where the ether will be forwarded
    - _rate: The rate at which the tokens should be minted
    - _weiRaised: Total wei raised in the crowdsale
- **Events**:
    - TokensPurchased: This event is emitted when an investor purchases the token.
- **Constructor**:
    - This takes three arguments for the crowdsale—a wallet address to forward the ether, the uint256 rate and the ERC20 token address.
- **Functions**:
    - function (): The fallback function is payable in order to receive the ether in the contract, which then calls the buyToken() function to start the token purchasing process.
    - token(): This is a public view function that returns the address of the ERC20 token.
    - wallet(): This is a public view function that returns the address of the wallet to which ethers are sent.
    - rate(): This is a public view function that returns the rate at which the tokens will be minted.
    - weiRaised(): This is a public view function that returns the total wei that's been raised so far in the crowdsale contract.
    - buyTokens(): This is a public payable function that is also called from the fallback function. This function also allows an investor to directly call the function and purchase the tokens.

- `_preValidatePurchase()`: This is an `internal view` function that allows the inherited contracts to override this function and add more sophisticated pre-validations, which are executed before token purchase in the contract.

- `_postValidatePurchase()`: This is an `internal view` function that allows the inherited contracts to override this function and add more sophisticated post-validations, which are executed after token purchase is done in the contract.

- `_deliverTokens()`: This is an `internal` function that safely transfers the given number of ERC20 tokens from the crowdsale contract to the investor. Other contracts can override this so that they can add their own way of doing the token sale.

- `_processPurchase()`: This is an `internal` function that's used to process the transfer of the tokens. Inherited contracts can override this to provide custom logic.

- `_updatePurchasingState()`: This is an `internal` function that updates the purchase state. Inherited contracts can override this to add custom logic; for example, to notify another registry contract about storing the information of the purchaser.

- `_getTokenAmount()`: This is an `internal` function that returns the number of tokens to be sent based on the wei that's sent. Inherited contracts can override this to provide their own custom logic in order to calculate the number of tokens to be generated for an investor.

- `_forwardFunds()`: This is an `internal` function that's used to forward the ether that's received in a transaction to the wallet address that has been provided in the constructor.

Create whitelisted crowdsale using WhitelistCrowdsale.sol

As we have seen, `Crowdsale.sol` is the basic version of the crowdsale contract and, on top of that, different features can be added. By using the `WhitelistCrowdsale.sol` contract, you can allow ether to be received from known/whitelisted addresses; other addresses cannot send ether to the contract. Whitelisted addresses can be investors' addresses, who are allowed to purchase tokens from the contract.

The OpenZeppelin contract present at `Chapter09/openzeppelin-solidity/contracts/crowdsale/WhitelistCrowdsale.sol` is defined as follows:

```
contract WhitelistCrowdsale is WhitelistedRole, Crowdsale {
    function _preValidatePurchase(address _beneficiary, uint256 _weiAmount)
    internal view {
        require(isWhitelisted(_beneficiary));
        super._preValidatePurchase(_beneficiary, _weiAmount);
    }
}
```

Let's review the preceding code in more detail. The contract contains the following function and inherits from the following contracts:

- **Inherits from**:
  - `WhitelistedRole`: A role for an address that has been whitelisted by `WhitelistAdminRole` previously
  - `Crowdsale`: The basic contract that supports crowdsale features
- **Functions**:
  - `_preValidatePurchase()`: This function pre-validates and allows only the whitelisted addresses that has `WhitelistedRole`

Other crowdsale contracts

Other types of crowdsale-related contracts are also present in the OpenZeppelin library. Let's look at where these contracts can be used:

- `validation/CappedCrowdsale.sol`: A crowdsale with an upper limit for the total wei or ether contribution also known as **hard-cap**
- `validation/IndividuallyCappedCrowdsale.sol`: A crowdsale with individually capped upper limit of wei investments
- `validation/PausableCrowdsale.sol`: Allows investment only when it isn't paused by `PauserRole`
- `validation/TimedCrowdsale.sol`: A crowdsale that opens and accept ether for a specified duration of time
- `distribution/FinalizableCrowdsale.sol`: Allows a special action to be triggered when crowdsale is over
- `distribution/PostDeliveryCrowdsale.sol`: A crowdsale that allows its investors to withdraw their tokens only after it finishes

- `emission/AllowanceCrowdsale.sol`: A crowdsale in which another wallet contains tokens and allowance is provided to the crowdsale contract so that it can deliver the tokens to the investors
- `emission/MintedCrowdsale.sol`: A crowdsale in which new tokens are minted only when investors send ether to the contract
- `price/IncreasingPriceCrowdsale.sol`: A crowdsale that increases the rate of the token linearly according to the time

You can also refer Chapter 12, *Building Your Own Token*, in which we have built our own crowdsale contract from scratch, using some of the above contracts.

Utility contracts

The OpenZeppelin library provides some utility contracts that can be used for general purpose functions, such as checking that an address is a contract, or a modifier to prevent reentrancy attacks.

The contract library files are in the `Chapter09/openzeppelin-solidity/contracts/utils` folder.

Check for contracts using Address.sol

In your contract, when you need to confirm whether an address is a contract or an EOA, you can use the `Address.sol` library. This library provides a function, `isContract()`, which checks the address and finds the byte size of the contract attached to that address. If the code size is greater than zero, that means it is a contract.

However, this is not the recommended way to check whether an address is a contract or not. During the contract's constructor execution, a contract isn't treated as a contract. Thus, if you call the `isContract()` function while a contract is executing its constructor, then the function would return `false`.

The OpenZeppelin contract present at `Chapter09/openzeppelin-solidity/contracts/utils/Address.sol` is defined as follows:

```
library Address {
    function isContract(address account) internal view returns (bool) {
        uint256 size;
        assembly { size := extcodesize(account) }
        return size > 0;
```

```
        }
    }
```

Let's review the preceding code in more detail. The contract contains the following library function:

- **Library function**:
  - `isContract()`: Finds whether a given address is a contract or an EOA address

Prevent reentrancy attacks using ReentrancyGuard.sol

There is a type of attack pattern where an attacker can call the function of a contract repeatedly. This is called a reentrancy attack. Using these types of attacks, an attacker may steal funds from the contract or perform unsafe operations on it. To prevent these types of attacks, the `ReentrancyGuard.sol` contract is used, which ensures that a function should be called only once per transaction.

However, it is the developer's responsibility to write code that isn't prone to reentrancy attacks. If there is still a chance of reentrancy attacks that cannot be stopped by code, then the modifier provided by this contract could be helpful.

The OpenZeppelin contract present at `Chapter09/openzeppelin-solidity/contracts/utils/ReentrancyGuard.sol` is defined as follows:

```
contract ReentrancyGuard {
    uint256 private _guardCounter;

    constructor () internal {
        _guardCounter = 1;
    }

    modifier nonReentrant() {
        _guardCounter += 1;
        uint256 localCounter = _guardCounter;
        _;
        require(localCounter == _guardCounter);
    }
}
```

Let's review the preceding code in more detail. The contract contains the following state variable, constructor, and modifier:

- **State variable**:
  - `_guardCounter`: This is an `internal` counter variable that increases per successive calls.
- **Constructor**:
  - In the constructor, the `_guardCounter` variable is initialized and set to `1` as a value.
- **Modifier**:
  - `nonReentrant`: This modifier can be used in a function to guard that function against reentrancy attacks.

You can refer `Chapter 14`, *Tips, Tricks and Security Best Practices, Reentrancy attacks*, for more details on reentrancy attacks.

Summary

In this chapter, we looked into some of the library contract files that are provided by OpenZeppelin that are commonly used during contract development. We learned how to maintain ownership in the contract using `Ownable` and `Claimable` contracts. Then, we discussed how to manage roles such as `PauserRole` and `MinterRole`; we also discussed the `Pausable` life cycle contract. We also did a deep dive into ERC20 standard-specific library contracts that can help in creating an ERC20 token with different features, such as mintable, pausable, and burnable ERC20 tokens. Later, we learned about `Math`, `SafeMath`, `Crowdsale`, and other utility contracts that are available in the library.

In this chapter, we only discussed some of the OpenZeppelin contracts. However, there are many other contract files present in the library. You can explore those contract files and use them in your decentralized application if you wish.

As a developer, we should keep checking OpenZeppelin contract files for bug fixes and new releases since critical issues were found in the library files that were added and then removed from the OpenZeppelin project. One such example is the ERC827 token standard, which was removed from the library due to bugs. Not only that, but ERC20 contract-related files also had some bugs, which have been fixed now.

The contract files that are present in the `draft` folder of the library are just the proposed draft contracts. You shouldn't consider using these libraries in the first place because these contracts haven't undergone heavy security auditing and were not used in production. Hence, it is likely that there are some bugs in these contract files.

In the next chapter, we will learn about multisig wallets, which allow multiple accounts to sign a transaction in order to get it executed.

Questions

1. Can we just copy-paste the OpenZeppelin library code into our project?
2. Should we modify the OpenZeppelin library files according to our needs and use them in our project?
3. What should we do if we don't need a function in our contract that is defined in the OpenZeppelin library?
4. Can we perform a mathematical operation without using the `SafeMath.sol` library?
5. Which is preferred—`Ownable` or `Claimable`?

Using Multisig Wallets

Multisig wallets are a special type of contract that requires multiple EOAs to sign and initiate the transaction. These types of wallets are helpful in maintaining a large sum of ether or tokens in a wallet. Mostly, they are used by companies to keep their funds safe, and only move funds when the required number of owners come to a consensus and confirm and execute the transaction. Users can also transfer the administration rights of a contract to a multisig contract so that the owners of the multisig wallet can decide and execute a function of the owned contract.

In this chapter, we will learn about the ConsenSys multisig implementation and how it works. We will set up our own multisig wallet using the Gnosis DApp and learn how to control contracts with it.

In this chapter, we will discuss the following topics:

- Understanding multisig wallets
- The benefits of using multisig wallets
- Precautions when using multisig wallets
- Setting up your own multisig wallet
- Controlling other contracts with multisig wallets

Technical requirements

The code that will be used in this chapter can be found on GitHub at `https://github.com/PacktPublishing/Mastering-Blockchain-Programming-with-Solidity/tree/master/Chapter10`.

You'll find a submodule folder called `MultiSigWallet` in the `Chapter10` folder. If the `MultiSigWallet` folder is empty, run the following command under the `Chapter10` folder to update it:

```
git submodule update --init --force --recursive --remote
```

In this chapter, we will be using the Gnosis multisig DApp GitHub repository as a submodule, which is present in the `MultiSigWallet` folder. The Gnosis repository uses the multisig Solidity contract that was written by ConsenSys. The project is old and hasn't been upgraded to use the latest version of Solidity. Therefore, the following Solidity version and tools are required for this project:

- Solidity version 0.4.15
- Truffle version 3.4.11 or higher

Understanding multisig wallets

As the name suggests, a multisig wallet is a wallet that requires multiple signatures to execute a transaction. A multisig wallet can have two or more owners controlling the transactions and funds that are present in the wallet. In a multisig wallet, you can define the minimum number of signatures that are required to execute a transaction. It's only when the required number of EOAs sign and confirm a particular transaction that it gets executed on the multisig. Existing owners can add or remove a new or existing owner from a multisig.

Different multisig wallet implementations are available, such as ConsenSys and Parity. However, the most widely used and safe multisig is the ConsenSys multisig implementation. In this chapter, we are going to talk about the ConsenSys multisig wallet implementation. The topics that we will discuss in this chapter will focus on, and are related to, the implementation of ConsenSys. There are different approaches to working with other multisig implementations, but we aren't going to discuss these here.

As we have already discussed in previous chapters, in Ethereum, there are two types of accounts—EOA and contract accounts. An EOA account is controlled using its private key. However, a contract account (smart contract) might have some functions that are only executed by an authorized EOA or contract. A multisig wallet is a smart contract in which multiple EOA accounts are authorized to control it. These EOA accounts are also known as parties or owners and have transaction signing rights so that they can confirm and execute transactions. Once the required number of confirmations are received for a transaction, then the transaction is executed.

The following diagram is an example of a multisig contract that's been deployed by a company to hold ETH and ERC20 tokens. As you can see, the contract has four owners, and three signatures are required to execute a transaction:

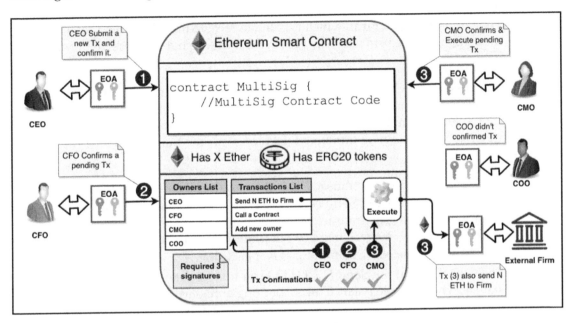

Multisig wallet of a company that keeps ETH and ERC20 tokens

In the preceding diagram, a company is sending **N** number of ethers to an external firm from its own multisig wallet. Let's understand how this works.

In the preceding diagram, we have the following entities and conditions:

- A company has a multisig contract deployed on the Ethereum blockchain:
  - The contract is using the ConsenSys multisig implementation as its contract code.
  - This contract has X number of ethers as well as some ERC20 tokens.
  - The multisig maintains a **Transactions List** where it keeps all the transactions and their corresponding confirmations.
  - The multisig maintains an **Owners List**, which represents the currently authorized signatories of the multisig.
  - The multisig requires at least three confirmations to execute a transaction.

- There are four owners of this multisig: the company's **CEO**, **CFO**, **CMO**, and **COO**. Each one has their own EOA account that's authorized to control the multisig. In other words, all of them are authorized to sign the transaction on multisig.
- There is an **External Firm**, which is going to receive the ether from the multisig of the company. The firm also has its own EOA in which it will receive the ether payment.

We can assume that each EOA transacting with the multisig has sufficient ether to perform a transaction. We can also assume that the multisig has a sufficient ether balance to send an ether payment to the **External Firm**. Let's understand the transaction sequence shown in the preceding diagram. The following numbered list represents the transaction sequence numbers in the diagram:

1. The **CEO** sends the first transaction (shown as ❶ in the preceding diagram) to the multisig in order to add **Send N ETH to Firm**, which gets added to the **Transactions List**. The transaction's confirmation is given to the transaction, which is shown as a check mark (✔) in front of **Tx Confirmations** for CEO. As of now, we have one confirmation for this transaction.

2. The **CFO** sees that the **Send N ETH to Firm** transaction is in a pending state. CFO checks the balance for accounting and, once approved, send the second transaction (shown as ❷ in the preceding diagram) to confirm that pending transaction. Their transaction confirmation is updated in the multisig wallet, which is shown as a check mark (✔) in front of **Tx Confirmations** for CFO. Now, we have two confirmations for this transaction.

3. The **CMO** sends a third transaction (shown as ❸ in the preceding diagram) to confirm the pending transaction. Their transaction confirmation is updated in the multisig wallet, which is shown as a check mark (✔) in front of **Tx Confirmations** for CMO. However, after this has been confirmed, the multisig finds that the minimum required number of confirmations (three, in this case) has been reached. Due to this, it executes the pending transaction as a part of transaction ❸. As you can see, the multisig executes the transaction and sends N number of ETH to **External Firm** from the multisig's ether balance. Since this third transaction executes the pending transaction as well, the **CMO** pays the required transaction gas in ether in order to execute the transaction.

Now that we've learned how multisig wallets work, let's discuss the benefits of using one.

Benefits of using multisig wallets

There are multiple benefits of using multisig wallets, such as when you need more safety for your funds and access rights.

Multisig wallets are mostly used for securing funds and contract access rights. Let's understand their benefits:

- **Multiple owners**: As we have seen, in the case of a single account handling funds/access rights, it is possible for private keys to be compromised easily. This could become a single point of failure for a system. To increase the level of security, the funds/access rights should be given to multiple entities so that the risk can be distributed among different stakeholders. The owners must ensure that their private keys aren't stored on a public place or on a public machine; this would be a single point of failure if that machine is ever compromised.
  You can also configure the number of required signatures that a transaction executes. The key stakeholder of the multisig must ensure that access to the multisig is only given to trusted entities since untrusted/less-trusted entities or people in the multisig wallet could misuse it.

- **Fund safety**: Keeping ETH, ERC20 tokens, or any other ERC standard tokens in multisig wallets greatly improves the security of funds. The multisig wallet should be used when a considerably large amount of funds are to be maintained and there are multiple stakeholders. Using a multisig for small funds may not be beneficial for a user since multisig requires additional transactions to be executed, such as the submission of a transaction, confirming a pending transaction, or executing a transaction, and each transaction consumes transaction fees in ether.

- **Ownership role management**: Multisig wallets can be used to control the access rights of other contracts. Once the ownership or access rights role is transferred to the multisig contract, that role can also be maintained by the multisig owners. If your decentralized application's contract has a critical role that can be misused if it's compromised, then, ownership rights should be maintained using multisig wallets.

Precautions when using multisig wallets

Multisig wallets increase the safety of funds/access rights. However, there are some precautions we should keep in mind while deploying multisig.

When you deploy and initialize a new multisig wallet, you have to provide the number of required signatures to execute a transaction. For example, if you have four key stakeholders, you might require at least three to sign the transaction and execute it. This is a nice way to manage multisig access because even if an owner loses their private key, it is possible for the other three existing owners to replace the owner who lost the private key with a new owner address. However, if you've configured four owners and require all four to sign the transactions, your multisig wallet could get locked. This is a dangerous way of setting up a multisig wallet, and so this should be avoided.

Before setting up the multisig wallet, you need to assess the risks that are involved. You should have a risk management policy in place to handle these kinds of situations. It is always better to have one or more backup keys added to the multisig and keep the private keys of those accounts in a safe vault. By doing this, your contract will be safe against the issues we mentioned in the preceding example. Here, out of the four owners of the multisig, you would need at least three of them to sign a transaction so that it can be executed. In this situation, it would be better to add one backup owner and keep the private key of this account in a safe vault. Here, it is more important that the safety of the vault is also ensured by the stakeholder, otherwise someone who is untrustworthy could get hold of it and harm your contract or funds.

This is just an example so that you know how to reduce the risk of using a multisig wallet. However, it is up to the company or the key stakeholders to define the risk vectors and take action accordingly.

The same precautions must be taken when you're adding/removing a new/old owner from a multisig wallet.

Now, let's move on and understand the ConsenSys multisig contract implementation.

Learning ConsenSys multisig implementation

In the *Technical requirements* section, we discussed that we will be using the ConsenSys multisig contract implementation, which is written by the ConsenSys software company. However, the DApp client for this contract is written and maintained by Gnosis. Let's understand some of the important functions and features of the ConsenSys implementation contract, which is called `MultiSigWallet`.

The `MultiSigWallet` contract is used by many projects on mainnet since it has been security audited multiple times and has no vulnerabilities at the time of writing.

In the `Chapter10/MultiSigWallet/contracts` folder, you will find all the contracts related to the multisig wallet. You can also find the contract files at `https://github.com/gnosis/MultiSigWallet/tree/ca981359cf5acd6a1f9db18e44777e45027df5e0/contracts`. This repository contains the following contract files:

- `MultiSigWallet`: This is the main ConsenSys multisig implementation contract file.
- `MultiSigWalletFactory`: This is a factory contract that's used to deploy a new instance of the `MultiSigWallet` contract.
- `MultiSigWalletWithDailyLimit`: This is an extension of the `MultiSigWallet` contract. It can apply a limit of 24 hours on the amount of ether that can be transferred from the contract.
- `MultiSigWalletWithDailyLimitFactory`: This is a factory contract that deploys a new instance of the `MultiSigWalletWithDailyLimit` contract.

Let's understand the code of the `MultiSigWallet` contract. We won't go into the details of each function—you can refer to the code to understand how the contract works:

- **Struct**:
  - `Transaction`: This struct keeps hold of transaction metadata. This struct contains the following fields:
    - `destination`: Stores the `address` value of the destination.
    - `value`: Stores the value/amount to be sent to the destination as `uint256`.
    - `data`: Stores the data to be executed on the destination contract in `bytes`.
    - `executed`: Stores the executed status as `bool`.
- **State variables**:
  - `transactions`: A mapping of the transaction ID to the transaction object.
  - `confirmations`: A map of a map to store who has confirmed a particular transaction ID. This keeps a transaction ID mapped to a map of `address` and its Boolean flag, which represents confirmation.
  - `isOwner`: A map to keep the owner status of each signatory of the multisig wallet.
  - `owners`: An array to store the EOA addresses of each multisig wallet owner.
  - `required`: The number of required signatures for a transaction to execute.
  - `transactionCount`: The number of transactions that have been added to the contract.

- **Modifiers**:
  - `onlyWallet()`: Only allows the function call from the same multisig contract.
  - `ownerDoesNotExist()`: Allows the function call when a given owner doesn't already exist in the owner's map.
  - `ownerExists()`: Allows the function call when the given owner exists in the owner's map.
  - `transactionExists()`: Allows the function call when the given transaction is present in the `transactions` map.
  - `confirmed()`: Allows the function call when a transaction is confirmed by a given owner.
  - `notConfirmed()`: Allows the function call when a transaction isn't confirmed by a given owner.
  - `notExecuted()`: Allows the function call when a given transaction ID isn't executed.
  - `notNull()`: Checks that the given address is not `address(0)`.
  - `validRequirement()`: Validates the required parameter and ensures that the number of maximum signatories are always less than 50.
- **Constructor**:
  - The constructor of the multisig wallet takes the list of the owner's account addresses and the number of required signatures for a transaction to confirm.
- **Functions**:
  - `function()`: A fallback function that receives ether and triggers a `Deposit` event.
  - `addOwner()`: This function adds a new owner to the multisig wallet.
  - `removeOwner()`: This function removes the owner from the multisig wallet.
  - `replaceOwner()`: This function replaces an existing owner with a new one.
  - `changeRequirement()`: This function changes the required number of signatures.
  - `submitTransaction()`: This function submits a new transaction request to the multisig wallet. Any of the owners of the multisig can call this function and submit a new transaction. This submitted transaction will be executed once it has received the required number of confirmations.
  - `confirmTransaction()`: This function confirms an existing transaction has been initiated by another owner. Only existing owners of the multisig can confirm a pending transaction.

- revokeConfirmation(): This function revokes the confirmation status of a transaction that has been confirmed by the owner previously.
- executeTransaction(): This function executes a specific transaction ID, and only proceeds when the required number of owners have signed the transaction. This function is also executed as part of the confirmTransaction() function when the required number of confirmations have been received.
- isConfirmed(): This function checks whether the provided transaction ID has been confirmed.
- getConfirmationCount(): This function gets the number of confirmations that have been performed on a given transaction ID.
- getTransactionCount(): This function filters and gets the transaction count.
- getOwners(): This function gets a list of the multisig owners as an array.
- getConfirmations(): This function gets a list of owners' addresses who have confirmed the given transaction ID.
- getTransactionIds(): This function filters and returns the corresponding transaction IDs according to their transaction status.
- addTransaction(): This is an internal function that's used to add a new transaction.
- external_call(): This is a private function that's used to execute the function call on the destination address.

- **Events:**
  - Confirmation: An owner confirms an existing pending transaction.
  - Revocation: An owner revokes their confirmation from a function that hasn't been executed.
  - Submission: A new transaction is submitted to the multisig wallet and is to be executed after confirmation is received.
  - Execution: A pending transaction has executed successfully.
  - ExecutionFailure: A pending transaction has failed to execute.
  - Deposit: Ether is deposited into the contract.
  - OwnerAddition: This is used when a new owner is added or when an existing owner has been replaced.
  - OwnerRemoval: This is used when an existing owner is removed or when an existing owner has been replaced.
  - RequirementChange: The number of required signatures is changed.

Setting up your own multisig wallet

You can find the submodule for the multisig wallet contract code in this book's GitHub repository, in the `Chapter10/MultiSigWallet` directory. We are using the fork from the Gnosis multisig wallet repository. This repository may be updated with fixes and upgrades in the future.

In the MultiSigWallet project, Gnosis has built a DApp for the multisig contract that was built by ConsenSys. The smart contract was written by ConsenSys, and Gnosis built a good looking DApp for it. By using the DApp client, the wallet becomes very user-friendly, and anyone can create and use them. The only requirement for the multisig wallet DApp is that it needs MetaMask or a supported hardware wallet so that you can connect your account to it.

As we saw in the previous example, there were four signatories/owners of the multisig wallet. Each owner can install and run their own copy of the multisig wallet DApp on their machine so that they can sign the transaction from their own EOA. It's also possible for the company to deploy this DApp on a shared web server so that the company's employees can access the DApp and sign the transaction using their own EOA account.

You can also access the already deployed version of the multisig wallet DApp on `https://wallet.gnosis.pm`. However, it is recommended that you install your own instance of the DApp and use it for signing transactions.

To install and run the multisig wallet DApp, run the following commands under the `Chapter10` folder:

```
$ cd MultiSigWallet
$ npm install
$ npm start
```

If you find that the `MultiSigWallet submodule` folder is empty, and you are unable to run the preceding commands, then you can execute the following command, which will fetch the code from the `MultiSigWallet` GitHub repository:

```
git submodule update --init --force --recursive --remote
```

After using the `npm start` command, you will see the following output. The server will start on port 5000 on your machine IP address:

```
Starting up http-server, serving ./
Available on:
   http://127.0.0.1:5000
   http://192.168.1.6:5000
Hit CTRL-C to stop the server
```

When you open the `http://127.0.0.1:5000` URL in your browser, you will be asked to confirm the terms and conditions. Once you've done this, you should see the following dashboard. Only open this in the browser that you have MetaMask installed on since it will connect to it. As you can see, it is connected to MetaMask and shows the current ETH balance:

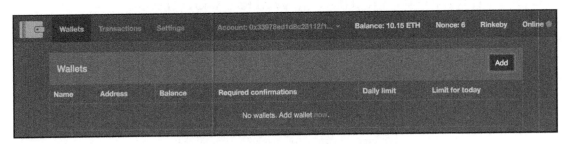

Gnosis multisig wallet DApp UI

In the preceding screenshot, we can see the **Wallets** section. We have no multisig wallets at the moment, which is why the list is empty. We will create our own multisig wallet in the next section.

Using the preceding multisig wallet DApp, you will be able to perform the following operations:

- Create and deploy a new instance of the multisig wallet contract. This allows the deployer to mention the number of owners and the required number of confirmations.
- Add and connect to an already deployed multisig wallet contract.
- Submit a new transaction via a multisig wallet, which is to be confirmed by other owners.
- Confirm pending transactions.
- Transfer ETH/ERC20 tokens from the contract.
- Add/remove/replace an existing/old owner from the multisig wallet.
- Connect using your Ledger or Trezor hardware wallets.

In this chapter, we will learn how to perform the following operations using the multisig wallet:

- Sending ETH from the multisig wallet
- Controlling contract authorization using a multisig wallet

Please note that we won't cover all of the features that are supported by this multisig wallet DApp/multisig wallet. We expect you to play around with this DApp and get acquainted with the other features.

Now, let's deploy a new multisig wallet and learn how to use it.

Deploying your multisig wallet contract

We will be using the Gnosis multisig DApp to perform all the required actions. For all of these transactions, we are using the Rinkeby testnet network. In the preceding screenshot, there's an **Add** button on the wallet dashboard screen. By clicking on this button, you will be asked to either **Create new wallet** or **Restore deployed wallet**:

Add wallet screen to create new or restore existing wallet

Since we are creating a new multisig wallet, we will choose the first option and click on the **Next** button. Here, you will be asked for the multisig wallet's metadata details, as shown in the following screenshot:

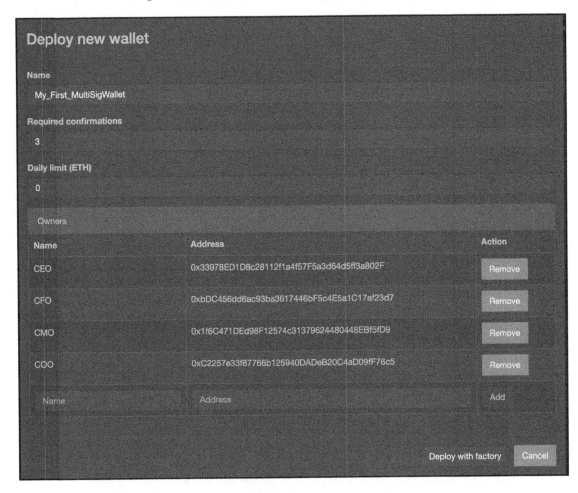

Deploying a new wallet screen

As you can see from the preceding screenshot, we have provided the following parameters:

- **Name**: The name for this multisig is `My_First_MultiSigWallet`. However, this name is only being used so that the DApp can maintain, identify, and manage multiple multisig wallets with ease. This isn't going to be set in the deployed multisig contract.
- **Required confirmations**: The number of minimum signatories that will be needed to sign a particular transaction to get it executed. Here, we set 3, since we want at least three out of four owners to be able to sign and confirm the transaction.
- **Daily limit (ETH)**: This is the maximum limit for the amount of ETH to be withdrawn/sent from this multisig contract on a daily basis. If this is set to 0, the limit is set to unlimited.
- **Owners**: A list of the owners' names and their respective EOA address. They will be the initial owners of this multisig wallet. The name field is only for the DApp to use, and name field isn't going to be set in the contract. However, EOA address will be set in the contract to maintain the owner's addresses. One thing to note here that an owner can also be an existing multisig contract address.

Once you have filled this in, click on the **Deploy with factory** button. You will then be asked for the transaction's gas parameters, as shown in the following screenshot:

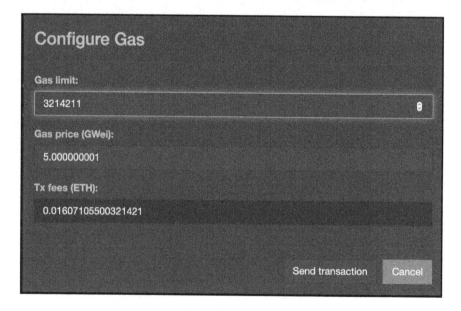

Configure gas screen to configure transaction gas parameters

As you can see, we are asked for the gas that's needed for this transaction to execute. Note that it will auto-calculate the **Gas limit** units for you, though this can be changed if you wish so. You can now click on the **Send transaction** button. When you are connected to MetaMask, you will be asked to confirm the transaction on the MetaMask screen popup:

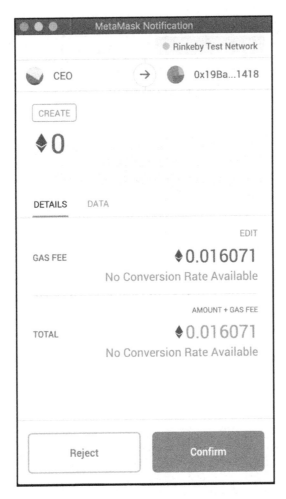

MetaMask transaction confirmation popup

As you can see, we are using the CEO's EOA account on MetaMask to deploy this contract. **CEO** is also one of the owner of the multisig wallet. Click on the **Confirm** button to confirm the transaction. Once the transaction has been executed successfully on the blockchain. you will see a new entry in the **Wallets** list, as shown in the following screenshot:

Multisig wallets list screen or Dashboard

As you can see, the new multisig contract has been created, and its address is **0xfD42883d303eEEC638DFAE9AEaE144075a6010B9**. You can also see the contract that it deployed on `etherscan.io`, as well as the contract code it published: `https://rinkeby.etherscan.io/address/0xfD42883d303eEEC638DFAE9AEaE144075a6010B9`. However, if you look at the contract name on etherscan, it will state that it has deployed the `MultiSigWalletWithDailyLimit` contract. This is the extension contract of the original `MultiSigWallet` contract and is used to support daily limits.

In the preceding screenshot, you can also see the current ETH balance of the multisig contract, which is currently **0.00 ETH**. You can also see that the **Required confirmations** is set to 3; you are allowed to change this setting if you so wish. Your contract's daily limit and limit for today are also shown in ETH.

Now, you have successfully deployed your first multisig wallet contract.

To view the latest transactions, you can click on the **Transactions** tab, which is present on the top bar of the screen:

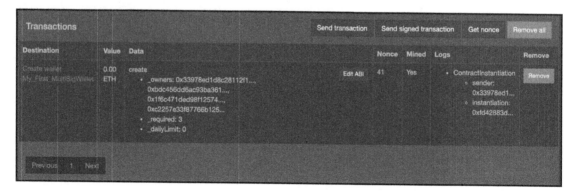

Transactions list screen showing history of transactions

As you can see in preceding screenshot, it lists all of the transactions that have been performed on the multisig wallet. There are also various buttons present on this screen. You can use these to send custom and signed transactions.

Now, let's have a look at how to send ethers from a multisig contract.

Sending ETH from a multisig contract

Now, let's learn how to send ETH from a multisig contract to any public EOA/contract address. We will replicate the example that we discussed in the first section of this chapter, *Understanding multisig wallets*.

We have deployed a multisig contract before. Initially, the multisig has zero ETH balance in it. First, we need to deposit some ETH into the multisig contract. To do this, click on the **Deposit** button, as shown in the following screenshot and follow the instructions to send ETH:

The multisig contract, its address, and its current ETH balance

You can also send ETH directly to your multisig contract address. Once you have deposited/sent ETH to the multisig contract, you will be able to see the available ETH balance, as shown in the preceding screenshot. Here, we have deposited **2.00 ETH** into the multisig contract.

Now that we have some ETH balance, we can initiate payments from the multisig contract to send ETH to another account. Remember that in order to send ETH from this multisig contract, you need three out of four owners to sign the transaction. We will assume that each owner has a sufficient ETH balance in their own EOA account to cover the transaction fees to initiate different types of transactions on the multisig.

Let's initiate a transaction and send **0.5 ETH** from the multisig contract to **External Firm X**.

Any of the four multisig owners can initiate a transaction. However, in this example, we are initiating the transaction from the CEO's EOA account. To initiate this, the CEO clicks on the `My_First_MultiSigWallet` contract link, which is shown in the preceding screenshot. This action will open another screen in which you will find a section called **Multisig transactions**, as shown in the following screenshot:

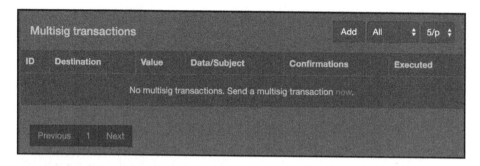

Multisig transactions screen

First, the CEO clicks on the **Add** button to initiate a new transaction. This action opens a new screen called **Send multisig transaction**, which is shown in the following screenshot. Here, you can directly put the **Destination** EOA account address to which you want to send ETH:

Address book button on the Send multisig transaction screen

However, we have already added an address to the address book. You can add an address to the address book by clicking on the **Address Book** menu in the top menu bar and providing an address, along with the associated name you would like to set. Since we have the address of **External Firm X** in our address book, we will click on the address book icon shown in the preceding screenshot. Clicking this opens the following **Browse address book** screen. Here, you can select the destination address that's present in the address book:

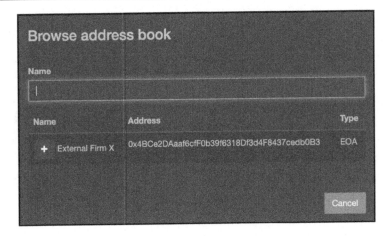

Browse address book screen

Here, we will click on the **+** sign present in front of **External Firm X** since we want to set this address as the **Destination** address so that we can send **0.5 ETH**. Upon clicking on this address, its information will be filled in on the **Send multisig transaction** screen, as shown in the following screenshot:

Send multisig transaction form filled in

Since we only want to send the **0.5 ETH**, we won't provide any other information. Just put 0.5 in the **Amount (ETH)** field, as shown in the preceding screenshot. Now, click on the **Send multisig transaction** button. This action will open the MetaMask popup screen, where you will be asked to confirm the transaction. Click on the **Confirm** button to do so:

CEO transaction initiated to transfer 0.5 ETH

Once the transaction has been mined, the DApp will auto-update the **Multisig transactions** section. In the preceding screenshot, you can see the following information:

- **ID**: The transaction ID of this multisig transaction is **0** (zero).
- **Destination**: This is the destination address on which the final transaction will be executed. Since this address is already present in our **Address Book**, it only shows the **Name** associated with that address.
- **Value**: This is the ETH value that will be sent once this transaction has been executed, after receiving the required number of confirmations.
- **Data/Subject**: This shows the transaction in plain English for better readability for the user. For this transaction, it's showing **Transfer 0.50 ETH to External Firm X**.
- **Confirmations**: This shows the current list of owners who have confirmed the transaction. As you can see, the **CEO** has already confirmed this transaction. Since MetaMask is using the CEO account, we can see that it also shows the **Revoke confirmation** button. This button is used to revoke the confirmation that's given to this transaction.
- **Executed**: This states whether this transaction has been executed or not. As we can see, this transaction is showing **No** as it hasn't been executed yet since it has only one confirmation. This transaction requires at least three confirmations before it can be executed.

Now, we need to change the account in MetaMask. We selected the CFO's account in MetaMask. However, in a real scenario, accounts won't be maintained in the same MetaMask instance or on the same machine. Each owner would have their own EOA account, either in MetaMask or in their Hardware wallet, and they would use their own Gnosis multisig DApp instance that's been installed on their machine.

By changing to the CFO's account in MetaMask, we can see that the DApp now shows the following:

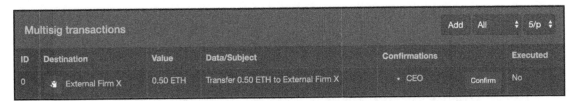

CFO's Multisig transactions screen, asking for confirmation

The **CFO** can see that transaction ID **0** hasn't been executed and that this is in a pending state. To confirm the transaction, the **CFO** needs to click on the **Confirm** button that's present on the preceding screen. By clicking this button, the MetaMask popup screen will open and ask for confirmation of the transaction from the CFO's account. You click on the **Confirm** button to confirm the transaction. Once the transaction has been mined and executed, you will see in the following screenshot that both the **CEO** and **CFO** have confirmed this transaction. Note that the **CFO** can revoke their confirmation from this transaction at any time.

Now, we will change the account in MetaMask and select the CMO's account. The **CMO** would see the following screen on their multisig instance:

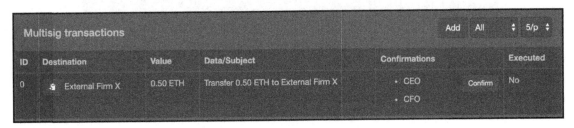

CMO's Multisig transactions screen, asking for confirmation

The **CMO** can see that transaction ID **0** hasn't been executed and that it is in a pending state. The transaction has been confirmed by **CEO** and **CFO**. **CMO** also click on the **Confirm** button to confirm the transaction from their side. This opens a MetaMask popup screen. Click on the **Confirm** button to confirm the transaction.

Once the transaction has been mined and executed, the CMO would see the following screenshot:

Confirmations from the CEO, CFO, and CMO, and the transaction executed

All of the owners of the multisig contract would be able to see the current status of the transaction **ID 0** in their own multisig DApp. As we can see in the preceding screenshot, the transaction has been executed. This means that the **0.50 ETH** has been sent from the multisig contract to an EOA address of the **External Firm X** account:

Multisig dashboard showing 1.50 ETH present

We can check whether the ETH has been sent by going to the multisig dashboard screen. To do this, click on the **Wallets** menu, which is present in the top menu bar. As shown in the preceding screenshot, the current balance of the multisig contract is **1.50 ETH**; earlier, it was **2.00 ETH**. Now, let's check the External Firm's EOA account in MetaMask:

MetaMask of External Firm X, which has an ETH balance of 0.5

As we can see, **External Firm X** has received **0.5 ETH**.

By doing this, we were able to successfully transfer **0.5 ETH** from our multisig wallet to an external EOA account of **External Firm X**.

Now, let's learn how to control the access rights of the contract using a multisig wallet.

Controlling contracts with multisig

In the previous chapters, we looked at different contracts that maintain access rights for a contract, such as `Ownable.sol`, `MinterRole.sol`, and `PauserRole.sol`. These access roles are basically given to an EOA. However, if the private key of that EOA account is stolen, then an attacker can potentially steal funds or perform malicious activities on the contracts.

If you transfer these kinds of access rights to an already deployed multisig address, more security will be provided. To do this, you can perform the following steps:

1. Deploy a new multisig wallet and initialize it with the required number of owners.
2. Ensure that the multisig wallet deployment is working and that the required number of owners are able to sign the transactions and execute it. This is done to ensure that the setup is working.
3. The ownership that you want to be maintained by the multisig contract should now be transferred to the multisig wallet address you deployed.
4. Your multisig setup is complete. Current owners should be able to confirm/execute transactions on multisig.

Using the Gnosis DApp of the multisig wallet, you can perform these types of transactions on the target contract using the ABI of the target contract.

Let's learn how you can control the access rights of a target contract using the multisig wallet contract.

Let's assume that we have our contract deployed and that some of the access rights (ownership or any role access) of the contract have been transferred to your company's multisig contract. Now, we can understand the transaction flow that's required to perform any action on the target contract. Take a look at the following diagram:

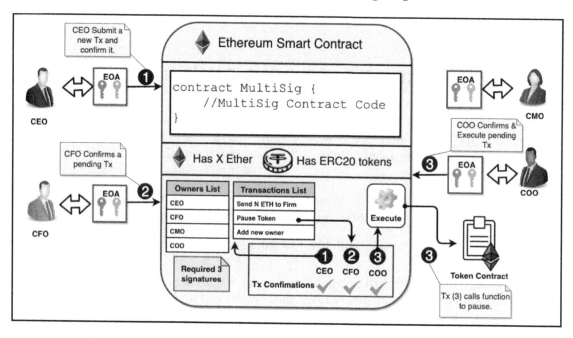

Multisig wallet controlling the PauserRole access right of a Token Contract

The preceding diagram is similar to the one we saw in the *Understanding multisig wallets* section of this chapter. The only difference is that the multisig here is controlling the `PauserRole` access right of a **Token Contract**. Let's understand the transaction flow from the multisig so that we can change the pause state of the **Token Contract**. Here, we will assume that the multisig contract has the `PauserRole` access right of a **Token Contract**:

1. The **CEO** initiates a transaction **Pause Token** so that they can pause the **Token Contract**. This transaction adds a new function execution transaction in the **Transaction List** and also confirms the transaction. This transaction flow is shown in the preceding diagram as ❶. The confirmation from the **CEO** is given, its shown as check mark (✔) in front of **Tx Confirmations** for CEO.

2. The **CFO** sees this transaction on their multisig wallet and confirms the transaction. The transaction flow is shown in the preceding diagram as ❷. The confirmation from the **CFO** is given, its shown as check mark (✔) in front of **Tx Confirmations** for **CFO**.

3. The **COO** sees this transaction on their multisig wallet and confirms the transaction. The transaction flow is shown in the preceding diagram as ❸. The confirmation from the **COO** is given, its shown as check mark (✔) in front of **Tx Confirmations** for **COO**. By doing this, the transaction is also executed on the target contract. This execution performs the Pause function call on the **Token Contract** in order to pause the contract. By doing this, no one is allowed to transfer their tokens using the **Token Contract**.

Let's look at an example of how you can maintain access rights using a multisig and control them. We are going to use the same example we have seen in the preceding diagram. To do this, we need to build a sample token called MyPausableERC20, as follows:

```solidity
pragma solidity ^0.5.0;

import
"https://github.com/OpenZeppelin/openzeppelin-solidity/contracts/token/ERC2
0/ERC20Pausable.sol";
import
"https://github.com/OpenZeppelin/openzeppelin-solidity/contracts/token/ERC2
0/ERC20Detailed.sol";

contract MyPausableERC20 is ERC20Pausable, ERC20Detailed {

    constructor() public ERC20Detailed("My Token", "SYM", 18) {
        //Mints 1 million token
        _mint(msg.sender, 1000000 * (10 ** 18));
    }
}
```

In the preceding code, we have a token with the following features:

- The token has the following metadata:
  - Name: My Token.
  - Symbol: SYM.
  - Decimals: 18.

- Once the token is deployed, it will mint **1 million SYM** tokens for its deployer address.

- It uses OpenZeppelin's `ERC20Pausable` contract to allow a `PauserRole` role to pause the contract.
- It uses OpenZeppelin's `ERC20Detailed` contract to allow it to specify the token metadata, such as `name`, `symbol`, and `decimals`.
- The deployer of the contract will be assigned `PauserRole`, which allows them to pause the contract at any time. Once the token is in the paused state, token transfers will not be allowed. However, token transfers will only work when contract is in the unpaused state.

To deploy this contract, paste the preceding code into Remix and use Solidity compiler version 0.5.0. Compile the contract and deploy it. You can refer to Chapter 4, *Learning MetaMask and Remix*, to learn how to deploy and access a contract using Remix and MetaMask.

We can also perform the following steps to deploy and transfer the `PauserRole` access rights of the token contract to a multisig contract. We are using the Rinkeby testnet to perform these steps:

1. You can use any account to deploy the `MyPausableERC20` token contract. However, we assume that the CTO of the company is setting up the contracts. We have deployed this contract here: `https://rinkeby.etherscan.io/address/` `0xad6430a99459c6593e401b36f4cdd2e91131d067`.

2. Once the contract has been deployed successfully, the CTO adds the `PauserRole` role for the multisig contract by calling the `addPauser()` function on the token contract and providing a multisig address as the function argument. The reference for the transaction we performed is `https://rinkeby.etherscan.` `io/tx/` `0x7585b881518347b3c1ae67b6fcd2260d9047d6a806e3925932736ee8275a8a00`. The multisig contract we are using here is the same one that we deployed in the *Setting up your own multisig wallet* section.

3. Now that the `PauserRole` role has been added, there's no reason for the CTO to have a `PauserRole`. Therefore, CTO call the `renouncePauser()` function to renounce the `PauserRole` access rights. Now, only the multisig has the `PauserRole` assigned. You can see the transaction details at `https://rinkeby.` `etherscan.io/tx/` `0x7721a2ed6ba314215f9484e0b10c9e4f04a4eb8335497a3f4b3cab153961987f`.

4. The CTO also transfers all of the SYM tokens (1 million) to the multisig contract. You can see the transaction details at `https://rinkeby.etherscan.io/tx/` `0x08dbeedfa0d8793b316d91357c667cb84d86171f30c6297fb85486b0e1ec3e6c`.

To check that the entire setup is correct, we can call the `isPauser()` function on the token contract and provide a multisig address as a function argument in Remix:

The isPauser() function call token contract

We can see that the multisig has the `PauserRole` assigned since the function call returned `true`. You can also check that the CTO doesn't have the `PauserRole` role; it should return `false`.

We can also check the current pause state of the `MyPausableERC20` contract. We do this by calling the `paused()` function in Remix:

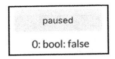

MyPausableERC20 token contract is in unpaused state

As we can see, the function returned `false`, which means that the function isn't paused and allows us to execute token transfers.

Now, we need to check the token balance in the multisig contract. To check the balance of any ERC20 token in your multisig DApp, you need to add the address of the token to the **Tokens** section. You can do this by going into the multisig, clicking on the **Add** button, and filling in the token address in the **Address** field. Wait for it to fetch the token metadata from the contract itself and show you its name, symbol, and decimals. Once you have confirmed these, click on the **OK** button. You will see the following screenshot:

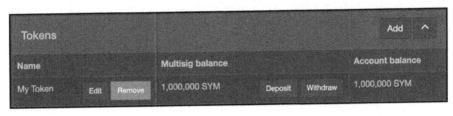

Token balance shows 1 million SYM tokens in the multisig

As you can see, the multisig contract has 1 million SYM tokens. You can also deposit or withdraw these tokens from the contract by clicking on the **Deposit** or **Withdraw** button, respectively. However, anyone can deposit tokens into multisig. They can do this by sending tokens to the multisig address. When you want to withdraw tokens from the multisig, the required number of owners (three, in this case) have to confirm the transaction, just like when we transferred ETH from our multisig to an address.

Everything has been set up: only the multisig contract has `PauserRole` access rights, and it also has 1 million SYM tokens.

Now, let's make a `pause()` function call from the multisig to the `MyPausableERC20` contract to pause the contract.

First, we need to update the **Address book** in the multisig DApp. Add the `MyPausableERC20` token contract address to the **Address book**, as shown in the following screenshot:

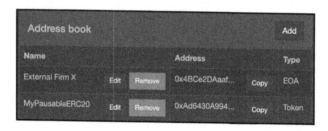

Address book maintained in multisig DApp

As we can see, it automatically detected the fact that **External Firm X** is the EOA address and that `MyPausableERC20` is a **Token**.

To perform any function call from the multisig to any contract, you need to have the ABI of the contract. The multisig DApp needs that ABI and finds the function calls that are allowed on a contract. To get the ABI of the contract from Remix, click on the **Compile** tab, select the `MyPausableERC20` contract from the drop-down list, and then click on the **ABI** button, as shown in the following screenshot:

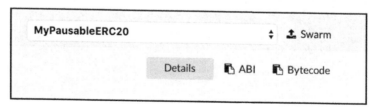

Remix IDE - Compile tab, MyPausableERC20 ABI, and the Bytecode button

Clicking the **ABI** button will copy the ABI code of the contract to your machine's clipboard. Now, all you need to do is go to your multisig DApp, select the multisig from **Wallets**, click the **Add** button of the **Multisig transactions** section, and paste the ABI into the **ABI string** field, as shown in the following screenshot:

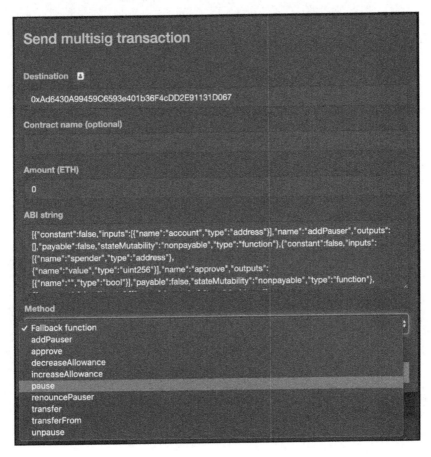

ABI function call using multisig

As you can see, we have selected the token contract address from the **Address book**, and it's been loaded in the **Destination** field. After loading the ABI in the **ABI string** field, it will auto-fill the **Method** dropdown, as shown in the preceding screenshot. Select the `pause` function from this list of functions. Then, click on the **Send multisig transaction** button. This will open a MetaMask popup, which will ask for transaction confirmation. Click **Confirm**. You will see that a new transaction has been added to the **Multisig transactions** list.

Now, select the **CFO**'s account from MetaMask. You will see the following transactions:

Multisig transactions listing the pause transaction on CFO's DApp

As we can see, the **CFO** can see that the `pause()` function call was initiated by the **CEO** to execute on the **My Token** contract. The **CFO** can confirm this transaction by clicking on the **Confirm** button. This will open a MetaMask popup that will ask for transaction confirmation. Click **Confirm**:

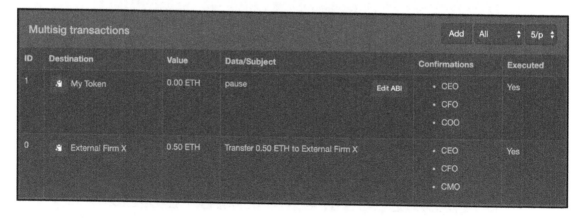

The pause() function confirmed by the CEO, CFO, and COO and its execution

The same process will also be followed by the **COO**. As you can see from the preceding screenshot, the three owners, **CEO**, **CFO**, and **COO**, have confirmed the transaction, and so it is executed.

To confirm that the token has been paused, go to Remix and call the `paused()` function:

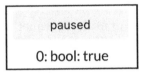

MyPausableERC20 Token has been paused

As you can see, the `MyPausableERC20` token has been paused. Now, token transfers won't be allowed by the token contract.

By doing this, you can call any function on any target contract on which multisig is authorized to call a function.

In the *Understanding multisig wallet* and *Sending ETH from a multisig contract* sections of this chapter, you saw how we can send ETH from a multisig. In this section, we have seen how we can control the access rights of a target contract and control it using multisig. You don't have to deploy multiple instances of a multisig to perform these kinds of transactions. You can maintain multiple roles and access rights even with a single multisig contract. This contract can also have ETH, ERC20, ERC721, or any other types of tokens as well. Therefore, by using a single multisig contract, you can control everything you would need, such as funds and access control.

Summary

In this chapter, we looked into one of the most well-known and battle-tested multisig: the ConsenSys multisig wallet. We learned how multiple owners can control a multisig contract and send ETH from a contract to another account. We also looked into how you can allow a multisig to maintain ownership or the role of a target contract. Every transaction on a multisig needs to have multiple signatures in order to be executed, which improves the security of the contract or funds. There are a few other multisig wallets available, but, at the time of writing, the ConsenSys multisig wallet seems to be one of the safest.

We also understood that, along with the fund's safety, multisig wallets can also be used for maintaining ownership access rights. These improve the multi-level security of your funds and access controls. It is always recommended to use multisig wallets for the critical access rights of contracts so that trust issues within the company can be mitigated.

Finally, we discussed the precautions to take while using multisig wallets. If they aren't configured properly, it could result in fund/access control loss.

Gnosis is also working on a new type of multisig wallet with new features, known as Gnosis Safe. However, at the time of writing, the project is still in its beta phase.

In the next chapter, we will look into the ZeppelinOS framework, which helps in creating and maintaining upgradeable contracts.

Questions

1. There are different implementations of multisig wallets present—which one should be used?
2. Can multisig wallets store assets other than ETH/ERC20 tokens?
3. Can a multisig wallet control another multisig wallet?
4. Is it possible to create a multisig wallet so that one of the owners from the owner's list has to sign every transaction?
5. What can we do if an owner of the multisig loses their private key?
6. How do we get an email notification regarding the transactions that are happening on a multisig?
7. What is the offline signing of a transaction?

11
Upgradable Contracts Using ZeppelinOS

In this chapter, we are going to learn about the ZeppelinOS development platform, which is used to develop, deploy, and operate smart contracts. This platform is like the Truffle framework but is used especially for upgradable contracts. Currently, there is no support in Ethereum for having an upgradable contract because it is a challenge for decentralization. However, you can write upgradable contracts on the Ethereum blockchain using the ZeppelinOS platform.

You will learn about the following topics in this chapter:

- Why there is a need for upgradable contracts
- Introduction to the ZeppelinOS platform
- ZeppelinOS installation and project creation
- Deploying contracts using ZeppelinOS
- Writing and upgrading existing contracts using ZeppelinOS

Technical requirements

The Solidity contracts, Truffle, and ZeppelinOS-specific files can be found on GitHub: `https://github.com/PacktPublishing/Mastering-Blockchain-Programming-with-Solidity/tree/master/Chapter11`.

For this chapter, you will need to install the following tools and frameworks as and when needed:

- Node.js v8.9.4 or later
- Truffle v5.0.4 or later
- Ganache GUI v2.0.0 or later/Ganache CLI v6.2.5 or later
- `zos` v2.3.1 – the GitHub location of the project is `https://github.com/zeppelinos/zos`
- `zos-lib` v2.3.1 – the GitHub location of the project is `https://github.com/zeppelinos/zos/tree/master/packages/lib`
- `openzeppelin-eth` v2.1.3 or later – the GitHub location of the project is `https://github.com/OpenZeppelin/openzeppelin-eth`

Understanding upgradable contracts

In `Chapter 1`, *Introduction to Blockchain*, we discussed how, on the Ethereum blockchain, once a Solidity smart contract is deployed, it cannot be changed. Because smart contracts are immutable, only the states of the variables can be changed by an authorized person. We also discussed how, because of the immutability of contract code, it is very important to not have any bugs left in contract code, since this could cause issues later, once the code is deployed to production.

Because of the immutability of contracts, there have been many attacks happened in the past on smart contracts. Once an attacker identifies bugs in contracts with funds, they exploit the bugs and vulnerabilities present in the contract and steal the funds. One famous hack was the Parity MultiSig hack, in which an attacker was able to become the owner of the contract and steal funds kept in multisig contract. Another hack related to Parity MultiSig was when a shared Solidity library was killed by someone which locked millions of dollars' worth of ether into multisig contract, with no possibility of recovering those funds as they are locked forever.

On one hand, the immutability of smart contracts plays an important role in decentralization. On the other hand, bug fixing or applying any patch to a contract is not possible. With this in mind, people started to think about how to write a contract that could be upgraded if any bug or vulnerability was discovered, so the funds present in the contract could be secured.

There is no direct support offered by Solidity or the **Ethereum Virtual Machine (EVM)** for upgrading a smart contract because, for true decentralization, smart contracts should not be allowed to change once deployed. However, people have found a way to write upgradable contracts using proxy contracts.

It cannot be said that you can build purely decentralized applications by using upgradable contracts, since those contracts can be upgraded by an authorized person. Hence, there is always centralized control of a contract and an authorized person can change the contract definition at any time too.

Let's look at the ZeppelinOS platform, which can be used to make a contract upgradable so that, in the future, new code can be added to fix bugs or add a new feature to the contract.

Introduction to ZeppelinOS

ZeppelinOS is a development platform used to develop, deploy, and manage upgradable contracts. The platform uses the Truffle framework, which is built especially for managing upgradable contracts.

As we discussed earlier, there is no native way to write an upgradable contract on the Ethereum blockchain, hence developers are left with some patterns with which they can make contracts upgradable. One such design pattern is the proxy design pattern, in which a contract behaves like a proxy to forward function calls to the target contracts. The administrator of the proxy contract can change the target contract; however, one limitation of the proxy design pattern is that the contract API cannot be changed. The function API that was being supported by the proxy contract is only able to forward the call to the target contract because, once it has been deployed, the proxy contract's code cannot be changed. That proxy contract is called a static proxy, as only predefined function calls can be forwarded to the implementation contract.

The following call flow diagram shows a transaction via a **Proxy Contract** to its current implementation contract version:

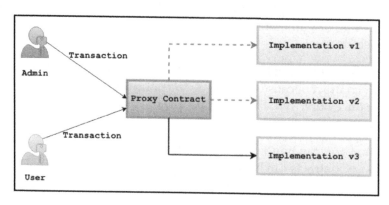

Proxy contract architecture and transaction flow

In the preceding diagram, the **Admin** has the privileges to change the implementation to be used in the **Proxy Contract**. Once the implementation contract address is changed, all transaction calls from the **User** will be forwarded to the new implementation contract. The solid line from the **Proxy Contract** to **Implementation v3** signifies that the proxy is currently configured to use **Implementation v3**.

In the static proxy contract, APIs are defined for each individual function present in implementation contracts, so that the **User** can call these API functions on the **Proxy Contract**. However, with this pattern, the storage of the state variables and function executions are performed at the implementation contract level. The **Proxy Contract** only has the implementation contract address stored; it only forwards the function call. The main drawback of this static proxy is that, if you wanted to add new functionality to be implemented, it would require the redeployment of a proxy as well, which would break the proxy pattern.

However, when creating upgradable contracts using ZeppelinOS, the Proxy contract stores all the state variables and execution is performed at the Proxy contract level. The execution code is referenced from the implementation contract only, but the code is executed at the Proxy contract level. This way, the Proxy contract becomes a dynamic proxy and can execute any function call. Let's look at how this is achieved at the Solidity level. The following is the code for the Proxy contract, which can handle dynamic function calls and forward them to implementation (target) contracts:

```
contract Proxy {

    function () payable external {
```

```
    _fallback();
  }

  function _implementation() internal view returns (address);

  function _delegate(address implementation) internal {
    assembly {
      calldatacopy(0, 0, calldatasize)
      let result := delegatecall(gas, implementation, 0,
      calldatasize, 0, 0)
      returndatacopy(0, 0, returndatasize)

      switch result
      case 0 { revert(0, returndatasize) }
      default { return(0, returndatasize) }
    }
  }

  function _willFallback() internal {
  }

  function _fallback() internal {
    _willFallback();
    _delegate(_implementation());
  }
}
```

We know that the fallback functions present in the contract will be executed when the EVM does not find a function definition present in the contract that has been requested in the transaction. As you can see in the preceding code, the fallback function simply forwards the call to the internal _fallback() function, which further forwards the call to the _delegate() function. The _delegate() function definition has assembly code that uses delegatecall for the contract implementation. This way, it can handle dynamic function calls in the Proxy contract, and execution of the function is performed at the Proxy contract level.

The preceding code used an assembly language block. This is called an inline assembly block containing assembly instructions. Assembly language instructions give you more fine-grained control over a contract. You can use assembly language when it is hard to achieve the required behavior from the contract. An upgradable contract is a special case where you need to access different memory locations.

Creating a ZeppelinOS project

Let's create a new project from scratch and use the ZeppelinOS development platform to write an upgradable contract:

1. Create the project folder:

    ```
    $ mkdir project
    $ cd project
    ```

2. Under the project folder, initialize the project with the npm package. Follow the instructions and provide the required parameter when asked by the command. This will create a package.json file:

    ```
    $ npm init
    ```

3. Install the ZeppelinOS development platform command and the related files in the current project. This will install a zos command that we can use further:

    ```
    $ npm install zos-lib@2.3.1 zos@2.3.1
    ```

4. Initialize the project with the zos command, as follows:

    ```
    $ npx zos init project v1
    ```

    This command will create two new files, called zos.json and truffle-config.js. The zos.json file is the configuration file for zos in which all the contract-related details are kept.

    The v1 parameter is the version name of the contract we are defining. The preceding command will create the zos.json file, which contains the following file contents:

    ```
    {
        "zosversion": "2.2",
        "name": "project",
        "version": "v1",
        "contracts": {}
    }
    ```

The content of the truffle-config.js file will be as follows:

```
module.exports = {
  networks: {
    local: {
      host: 'localhost',
```

```
      port: 9545,
      gas: 5000000,
      gasPrice: 5e9,
      network_id: '*',
    }
  }
}
```

As you can see, the network name is `local`, and Truffle would look to connect to Ganache on `localhost` and port `9545`.

The `zos-lib` library contains the library contracts related to contract upgradability.

Let's now create an upgradable contract, `StableToken.sol`:

```
import "zos-lib/contracts/Initializable.sol";
contract StableToken is Initializable {

    function initialize(
      address _owner,
      string memory _name,
      string memory _symbol,
      uint8 _decimals
    )
    public initializer
    {
      owner = _owner;
      name = _name;
      symbol = _symbol;
      decimals = _decimals;

      uint tokensToMint = 1000000 * (10 ** uint(decimals));
      totalSupply += tokensToMint;
      balances[owner] += tokensToMint;
    }

    //DO NOT USE THIS IN PRODUCTION
    //Integer overflow possible
    function transfer(address _to, uint _amount) public returns (bool) {
      balances[msg.sender] -= _amount;
      balances[_to] += _amount;
      return true;
    }
}
```

The actual code file is relatively big; hence, we have only shown part of the code. For the full code, you can head to this GitHub link: `https://github.com/PacktPublishing/Mastering-Blockchain-Programming-with-Solidity/blob/master/Chapter11/contracts/StableToken_v1.sol`. The code is kept in the `StableToken_v1.sol` file. You can use the code and put it in your newly created `project/contracts` folder.

The `StableToken` contract inherits from the `Initializable` contract. The `Initializable` contract is defined in the `zos-lib` package, which helps in initializing upgradable contracts. This contract has a `initializer` modifier, which we are using in the `StableToken` contract.

As you can see in the preceding code, no constructor has been defined. It is a requirement of an upgradable contract with ZeppelinOS that the contract must not have a constructor defined. Instead, it should be replaced with an `initialize()` function. As you can see, the `initialize()` function is using the `initializer` modifier. This ensures that the `initialize()` function can be called only once.

The preceding code is prone to integer overflow attacks; hence this code should not be used in production. We are only using this code to showcase upgradable contract creation and upgrade it with a fix. We intentionally created the contract with a bug so that we could upgrade it with a fix, using the `zos` framework.

The code mints 1 million tokens when the contract is initialized, and a token holder is allowed to transfer the tokens to another person.

Now, let's add the contract to `zos` so that it can be made upgradable with the `zos` framework. To do this, run the following command:

```
$ npx zos add StableToken
```

The preceding command will register the `StableToken` contract to the list of contracts that will be upgraded via `zos` and will also update the `zos.json` file. As we have added the contract in the `zos` configuration file, let's deploy it.

Deploying the StableToken contract

We are going to access the local blockchain using the `ganache-cli` command. If `ganache-cli` is not already installed on your machine, you can run the following command to install it; otherwise, you can ignore it and directly start the blockchain:

```
$ npm install -g ganache-cli
```

To start the local blockchain, follow these steps:

1. Open a new Terminal window and execute the following command to start the Ganache local blockchain instance on port 9545:

   ```
   $ ganache-cli -p 9545
   ```

2. Once the ganache-cli command has started the local blockchain instance, go back to the old Terminal window and run the following command:

   ```
   $ npx zos session --network local --from
   0x0421E4eB85b5a8443A8Fd5cc8ab3FfA0B723Db26 --expires 3600
   ```

In the preceding command for the --from parameter, we are using the public key of the very first wallet address used by the ganache-cli tool. You can find that address in the ganache-cli console output.

Once executed, zos will ask for a time-out in seconds for all web3 transactions, as follows:

```
? Enter a timeout to use for all web3 transactions (in seconds)
3600
```

Just press *Enter* to continue using a 3,600-second time-out for all web3 transactions.

The session command will start a session with the given --network option; here, we have chosen to use the local network. The --from parameter takes the wallet address from which the transactions will be sent, and the --expires parameter specifies the timeout for a transaction in seconds.

The preceding command will output the following in the Terminal window:

```
Using network development, sender address
0x0421E4eB85b5a8443A8Fd5cc8ab3FfA0B723Db26 by default.
```

3. Now, we can push the contract onto the blockchain using the following command:

   ```
   $ npx zos push
   ```

This command returns the following output:

```
Using session with network development, sender address
0x0421E4eB85b5a8443A8Fd5cc8ab3FfA0B723Db26, timeout 600 seconds
Validating contract StableToken
```

```
Uploading StableToken contract as StableToken
Deploying logic contract for StableToken
Created zos.dev-1557538865119.json
```

As you can see, the `push` command first compiles the contracts, deploys the `StableToken` contract, and returns its address. The `push` command deploys all the contracts that you previously added to `zos`.

As you can see in the previous result, the `push` command created the `zos.dev-1557538865119.json` file, where `1557538865119` is the blockchain network ID number. As we are running the local blockchain using `ganache-cli`, the network ID for that blockchain is random with every run; however, when using mainnet or testnet, the network ID is fixed. The generated file contains information related to the contract and its deployed address. If the file already exists, the information present in the file is updated with the new contract address on which the contract is deployed.

One thing to note here is that the contracts deployed using `push` are only the logic contracts. These contracts should not be used directly. Instead, they should be accessed via the `Proxy` contract.

4. Deploy the `Proxy` contract and initialize the implementation contract. To do this, you will need to run the following command:

```
$ npx zos create StableToken --init initialize --args
0x0421E4eB85b5a8443A8Fd5cc8ab3FfA0B723Db26,StableToken,STKN,18
```

As you can see in the preceding command, we are passing the argument to the initialize the function present in the `StableToken` contract. The first parameter in the `--args` flag provides the very first public key of the account generated with `ganache-cli`, as we want to make the first account the owner of the token. The command will generate the following output:

```
Deploying new ProxyAdmin...
Deployed ProxyAdmin at 0x316C3A2047438e7951566B48F4dA154281F12c84
Creating proxy to logic contract 0x664AA5b28b5c36D2622f09662eacf29635DAA105
and initializing by calling initialize with:
 - _owner (address): "0x0421E4eB85b5a8443A8Fd5cc8ab3FfA0B723Db26"
 - _name (string): "StableToken"
 - _symbol (string): "STKN"
 - _decimals (uint8): "18"
Instance created at 0xF2A348783dC30b8219854b647fA157448256deeE
0xF2A348783dC30b8219854b647fA157448256deeE
Updated zos.dev-1557538865119.json
```

In the preceding command output, you can see that the proxy instance for the StableToken contract is created at the 0xF2A348783dC30b8219854b647fA157448256deeE address. This command also updates the zos.dev-1557538865119.json file with the newly created proxy instance address. You can find the address of the proxy instance in the JSON file at proxies.project/StableToken.address. Once this contract instance is created, you should always interact with the implementation via the proxy instance. Never directly interact with the implementation logic contract address, because when you upgrade the contract implementation logic, the implementation contract will be deployed at a new address, although the proxy instance contract and its address remain the same.

You can also check the status of contracts and projects using the following command:

```
$ npx zos status
```

This command will return the following output with the current status of the contracts:

```
Project status for network dev-1557538865119
Application contracts:
- StableToken is deployed and up to date
Deployed proxies:
- project/StableToken at 0xF2A348783dC30b8219854b647fA157448256deeE version
v1
```

As you can see in the preceding result, the StableToken contract is deployed and up to date. The version of the contract is v1.

We have deployed the implementation logic contract and the Proxy contract for that implementation. Now, let's look into upgrading the contract after making changes to the contract code.

Upgrading the contract

In the previous section, we looked at the StableToken contract code, which is prone to integer overflow attacks. Once the bug is identified for a production contract, you can upgrade the implementation contract and deploy it to fix the issue.

ZeppelinOS manages the upgrading of the contract logic and storage structure seamlessly. Let's now use the following steps to fix the integer overflow issue present in the StableToken contract:

1. Manage the version of the project so that it becomes easy to track the changes that were introduced and fixed in a specific version of the contract. For that, let's bump the project version:

   ```
   $ npx zos bump v2
   ```

   This command will bump the project version from v1 to v2 and update the zos.json file accordingly.

2. Change the code to fix the issues:

   ```solidity
   contract StableToken is Initializable {
       //Introduced in version v2
       address public minter;
       event Transfer(address indexed _from, address indexed _to,
       uint256 _value);

       modifier onlyMinter() {
           require(msg.sender == minter);
           _;
       }

       //Fixed integer overflow issue in v2
       function transfer(address _to, uint _amount) public returns
       (bool) {
           require(balances[msg.sender] >= _amount);
           balances[msg.sender] -= _amount;
           balances[_to] += _amount;

           emit Transfer(msg.sender, _to, _amount);
           return true;
       }

       function initializeMinter(address _minter) public {
           require(minter == address(0));
           minter = _minter;
       }

       function mint(uint _newTokens) public onlyMinter {
           uint tempTotalSupply = totalSupply + _newTokens;
           require(tempTotalSupply >= totalSupply);

           totalSupply += _newTokens;
   ```

```
        balances[minter] += _newTokens;
    }
}
```

The actual code file is a bit large, so we have only shown part of the code. For the full code, you can head to this GitHub link: `https://github.com/PacktPublishing/Mastering-Blockchain-Programming-with-Solidity/blob/master/Chapter11/contracts/StableToken_v2.sol`. The code is kept in the `StableToken_v2.sol` file. You can now copy and paste this file into the `project/contracts` folder. Ensure that you remove the `StableToken_v1.sol` file from the folder, so that only a single `StableToken` contract definition is present in the `contracts` folder.

As you can see in the preceding code, we have fixed the integer overflow issue present in the `transfer()` function. We have also introduced an `address` `minter` variable to the contract, which has the permission to mint new tokens anytime via the `mint()` function. The `minter` user's address will be stored in a state variable—`minter`. Also, note that we have added the variable definition at the end of the preexisting state variables list. Later in this chapter, we will discuss the recommendations and precautions that need to be taken when adding new variables to the upgraded contract.

Also, note that a new function, `initializeMinter()`, has been added to initialize the `minter` state variable in the new contract version. We could have added a new parameter in the `initialize()` function itself; however, we cannot do this as the `v1` version contract is already initialized, and, to add a new state variable in the existing contract, we need to add its initialization function separately. If you are deploying the contract for the first time, then you could add the parameters in the `initialize()` function itself.

3. Deploy the new version of the `StableToken` contract, `v2`. However, one thing to note here is that, as of now, we are running and executing these contracts on the local blockchain, and to preserve the previous state of the blockchain, you must have kept the previous `ganache-cli` session running. In the case of testnet and mainnet, you do not have to worry about it. Our contract code is ready, so we can first push the contract that will deploy the `v2` version of the `StableToken` contract:

   ```
   $ npx zos push
   ```

This command would generate the following output:

```
Validating contract StableToken
- New variable 'address minter' was added in contract StableToken
in contracts/StableToken.sol:1 at the end of the contract.
See
https://docs.zeppelinos.org/docs/writing_contracts.html#modifying-y
our-contracts for more info.
Uploading StableToken contract as StableToken
Deploying logic contract for StableToken
Updated zos.dev-1557538865119.json
```

As you can see in the preceding command output, `zos` identifies that the new `address minter` variable is added into the contract; it also validates that the new state variable is correctly added at the end of the contract. Then, it proceeds with the deployment of the logic contract.

4. As the logic contract is deployed, let's update the `Proxy` contract with the new implementation and initialize the minter address by calling the `initializeMinter` function on the contract. In the following command, we are sending a second public key from the accounts generated with `ganache-cli`, as we want to give that account minter rights:

```
$ npx zos update StableToken --init initializeMinter --args
0x4cfae0c67f1b831d331934aec4cf437da6043bd5
```

The preceding command would generate the following output:

```
Using session with sender address
0x0421E4eB85b5a8443A8Fd5cc8ab3FfA0B723Db26, timeout 600 seconds
Upgrading proxy to logic contract
0x1E46054894464408F200eF48a2fc16a5f213E7Fa and initializing by calling
initializeMinter with:
 - _minter (address): "0x4cfae0c67f1b831d331934aec4cf437da6043bd5"
Upgrading proxy at 0xF2A348783dC30b8219854b647fA157448256deeE and calling
initializeMinter with:
 - _minter (address): "0x4cfae0c67f1b831d331934aec4cf437da6043bd5"...
TX receipt received:
0xec93c41d2008cddf845d658a6bda562f2c7c330ccc0a2d57642a09923d463092
Instance at 0xF2A348783dC30b8219854b647fA157448256deeE upgraded
0xF2A348783dC30b8219854b647fA157448256deeE
Updated zos.dev-1557590753547.json
```

As you can see, the address of the proxy remains the same
as 0xF2A348783dC30b8219854b647fA157448256deeE; however, the implementation
logic contract has changed. Also, we have added a new state variable, address minter,
which has also been initialized with the address passed in from the command.

Now that everything is done and updated, the proxy points to the new version,
the v2 implementation of the StableToken contract. To verify this, we can check the
current status as follows:

```
$ npx zos status
```

The command will generate the following output:

```
Project status for network dev-1557538865119
Application contracts:
- StableToken is deployed and up to date
Deployed proxies:
- project/StableToken at 0xF2A348783dC30b8219854b647fA157448256deeE version
v2
```

As you can see, the new version of the contract, v2, is being used by the proxy. As you can
see, the proxy address, 0xF2A348783dC30b8219854b647fA157448256deeE, remains the
same in version v1 and version v2.

We can also check that the new variables are properly initialized, and old variables remain
intact in the contract. To check this, we can use truffle console and run the following
command:

```
$ truffle console --network local
```

The command will open the truffle console Command Prompt as truffle(local)>,
under which you can run the following command. Make sure to use the deployed proxy
address in the following command:

```
truffle(local)> st = await
StableToken.at('0xF2A348783dC30b8219854b647fA157448256deeE')
```

The preceding command will initialize an internal variable, st, in Command Prompt,
which you can use to call the functions on the contract:

```
truffle(local)> st.name()
 'StableToken'
truffle(local)> st.owner()
 '0x0421E4eB85b5a8443A8Fd5cc8ab3FfA0B723Db26'
truffle(local)> st.minter()
 '0x4cFAE0c67F1b831D331934AEC4cF437dA6043bd5'
```

As you can see, all the existing variables from contract version v1 are intact, and the new minter state variable, added in version v2, returned the minter's address.

During the contract deployment and upgrade processes, we used some of the zos commands. However, there are many other commands that you can use according to your needs. Let's look at the zos supported commands.

ZeppelinOS commands

In the previous section, we used some of the useful zos commands. However, there are many more commands available in zos. Once the ZeppelinOS framework is installed, you can run the following command to check the installation and learn about the list of subcommands and options supported by the zos commands:

```
$ npx zos --help
```

The preceding command will generate the detailed output and list each command supported by zos. Let's discuss some of the most used zos commands and options:

```
Usage: zos <command> [options]
```

The following are the options:

- --version: Outputs the version of zos
- -v, --verbose: Runs a command in verbose mode to output errors, stack trace, and detailed logs
- -s, --silent: No output will be generated on the command line
- -h, --help: Outputs command usage information

The following are the commands:

- add [contractNames...]: Takes the contract names separated by whitespace characters and adds each contract to your project.
- bump <version>: Bumps your project to a new <version>.
- check [contract]: Checks your contracts for potential issues.

- `create <alias>`: Deploys a new instance of an upgradable contract for a given contract. You can provide an `<alias>` tag that you added your contract with, or `<package>/<alias>` to create a contract(s) using a linked package.

- `freeze`: Freezes the current release version of your published project.

- `init <project-name> [version]`: Initializes your ZeppelinOS project. Provides `<project-name>` and, optionally, an initial `[version]` name.

- `link [dependencies...]`: Links a project with a list of dependencies, each located in its npm package.

- `publish`: Publishes your project to the selected network.

- `push`: Deploys your project to the specified network.

- `remove [contracts...]`: Takes the contract names separated by whitespace characters and removes each contract from your project.

- `session`: Creates a session using the given network and wallet address. Some commands require a session before they are executed. These commands are `create`, `freeze`, `push`, `status`, and `update`. You can use `--close` to close a session.

- `set-admin [alias-or-address] [new-admin-address]`: To change the upgradability contract admin of a given contract. You can provide the `[alias]` or `[package]/[alias]` arguments of the given contract to change the admin ownership address of all its instances. You can also use the contract address (`[address]`) to change the admin of a single contract. Note that if you transfer contract ownership to an incorrect address that you don't have control of, you may lose control over upgrading your contract in future.

- `status`: Prints the information about the contracts, the address at which they are deployed, and the version and network information.

- `unlink [dependencies...]`: Takes dependency names separated by whitespace characters and unlinks each npm package dependency from the project.

- `update [alias-or-address]`: Updates the new contract logic on the network. You can provide the `[alias]`, `[package]/[alias]`, or `[address]` arguments you added your contract with. You can also use the `--all` flag to update all contracts preset in your project.

- `verify <contract-alias>`: This command verifies the given contract with Etherscan or Etherchain and provides the contract name.

Precautions while using ZeppelinOS

The ZeppelinOS development platform is built to help developers with maintaining upgradable contracts. As we discussed in the introduction section of this chapter, this upgradability is not a native concept in Ethereum, the EVM, or Solidity. It was developed using sophisticated Solidity upgradability design patterns, delegate calls, and assembly code. Hence, there are some recommendations that a developer must know about before working on writing upgradable contracts.

The state variables present in contracts are stored in storage slots. State variables take a position in the storage slots in the order in which they are defined in the contract. As upgradable contracts created using the zos platform, state variables are always stored in the Proxy contract; therefore, when the implementation changes, the state variables that are still present in the Proxy contract maintain the old state. Hence, the developer must know that changing the implementation code will affect the storage slot data present in a Proxy contract. If care is not taken when adding new state variables, it could corrupt the existing state variables while performing a contract upgrade.

Let's discuss the precautions that a developer has to take when adding, removing, or updating state variables.

Precautions for state variables

The state variables of a contract are stored in storage slots. When there is a collision between the storage slots used in the old version and the new upgraded contract, it could create problems and even the data stores in state variables may be corrupted. This could result in a serious loss or unintended behavior by the contract. Let's look at an individual technique to avoid a storage collision.

We assume that a current version of an example contract is defined as follows:

```
contract SampleContract {
    uint256 public x;
    string public y;
}
```

Now, we are going to upgrade this contract and we will look at different examples.

Avoid changing the variable declaration order

The order of the state variables present in the contract should not be changed. If you want to add a new state variable, add it at the end of all the state variables:

```
//Bad Practice
contract SampleContract {
    string public y;   // Variable order changed
    uint256 public x;
}
```

In the preceding code, the variable order has changed. Previously, the `uint256 public x` variable was defined first in the contract, and then it was the `string public y` variable. However, in the new code, the order has changed—now, the `string public y` variable is defined first. This should be avoided.

Avoid changing variable data types

The state variable data type must remain the same as the previous contract version. The data type should not be changed in the new contract version:

```
//Bad Practice
contract SampleContract {
    uint256 public x;
    bytes public y; //Data type changed
}
```

In the preceding code, the data type of the `y` variable has changed from `uint256` to `bytes`. This should be avoided.

Avoid adding new variables before existing variables

You should not add a new state variable before any of the existing state variables present in the contract:

```
//Bad Practice
contract SampleContract {
    uint256 public x;
    uint8 public decimals;   //Added new variable in between
    string public y;
}
```

In the preceding code, between the `x` and `y` variables, a new variable, `decimals`, has been added. This should be avoided.

Avoid removing existing state variables

You should not remove any existing state variables present in the old contract version. This should be avoided. For example, in the following code, the `string y` state variable has been removed:

```
//Bad Practice
contract SampleContract {
    uint256 public x;
    //string public y; //Variable removed / commented
}
```

In any case, if you were to add another variable after removing `string y`, the new variable would reflect the last value stored in the storage slot.

Always add new variables at the end

You should always add a new state variable at the end of all the existing state variables:

```
//Good Practice
contract SampleContract {
    uint256 public x;
    string public y;
    uint8 decimals;  // New variable added
}
```

In the preceding code, we have added a new state variable, `decimals`, at the end of the existing variables declaration.

Precautions when changing variable names

You should know that if you change the name of an existing variable, then the contract becomes highly error prone as follows:

```
// Should be done with caution
contract SampleContract {
    uint256 public count; // Variable name changed from 'x'
    string public y;
}
```

In the preceding case, the x variable has now been renamed `count`. However, when the contract is deployed, the `count` variable will still get the previous value assigned to the x variable. Hence, only the variable name is changed, but it will still give the old value.

Avoid initial values in the field declaration

In contracts, you can assign initial values to the state variable:

```
contract MyContract {
    uint256 public tokensToMint = 1000000;
}
```

As we know, in normal contracts, the tokensToMint variable will be initialized with the previously given value. However, when writing upgradable contracts using zos, you should not define initial values for the state variables in the contract. These values must be set in the initialize() function, as follows:

```
contract MyContract is Initializable {
    uint256 public tokensToMint;
    function initialize() initializer public {
        tokensToMint = 1000000;
    }
}
```

If these variables are not initialized in the initialize() function, then they will not be set.

Precautions for contract inheritance

Solidity does support contract inheritance. Hence, each of the base contracts can also have state variables present. Due to this, there could be the possibility of a storage collision, so you need to take care while writing an upgradable contract.

For example, let's take the following contract code:

```
contract A {
    uint256 a;
}

contract B {
    uint256 b;
}

contract MyContract is A, B {
    //...
}
```

The MyContract contract inherits from the A and B contracts. When the MyContract contract is deployed, the state variables present in the A and B contracts will get stored in storage slots. Let's understand the precautions that should be taken when using contract inheritance.

Avoid changing the inheritance order

If any change is made to the inheritance order, it will also affect the storage variable order. For example, say you were to change the inheritance in the following way:

```
//Bad Practice
contract MyContract is B, A { //Inheritance order is changed
    //...
}
```

This would update the compiled code of the contract and might corrupt the state variable data. Avoid doing this for contract upgrades.

Avoid adding new state variables in base contracts

You should avoid adding a new state variable in a contract that is being used as a base contract for another contract. In the following code, a new state variable, uint256 c, is added in the A contract. This would affect the storage slot layout when a contract is upgraded:

```
//Bad Practice
contract A {
    uint256 a;
    uint256 c; //New state variable added
}

contract B {
    uint256 b;
}

contract MyContract is A, B {
    //...
}
```

Hence, you should avoid adding new state variables in any base contracts. The way to solve this issue is to initially add some reserved state variables in base contracts, so that, in future, when there is a need, these reserved state variables can be used for upgraded contracts.

Summary

There is no direct native support for upgradable contracts in Ethereum. This is because pure decentralization cannot be achieved if a contract is upgradable. Once contract code becomes immutable, then only you can trust that the code and its logic will never be changed. However, due to invisible bugs present in contracts, a lot of hacking has taken place; therefore, ZeppelinOS came up with upgradable contracts to avoid bugs in contracts, even if they have already been deployed.

ZeppelinOS should not be used until there is a real need for it. You should write immutable contracts to have pure decentralization for your applications. If you plan to upgrade your contracts in future, then you should consider using ZeppelinOS.

In the next chapter, we will build our own ERC20 token from scratch. We will go through the process and design decisions, using the tools and libraries that are required to create an ERC20 token.

Questions

1. Should ZeppelinOS always be used?
2. Should we use the constructor for upgradable contracts?
3. How does ZeppelinOS affect decentralized trust models?
4. Is it possible to reinitialize a state variable?
5. Where is the implementation address stored, which is present in the `Proxy` contract?
6. Why should you not use implementation logic contracts directly to interact?
7. Can we remove any existing state variable?

12
Building Your Own Token

In previous chapters, we learned about Ethereum, Solidity, creating smart contracts from scratch, using MetaMask, Remix IDE, Truffle, and much more. In this chapter, we are going to use all the tools and techniques that we learned so far to create our own ERC20 standard token from scratch and sell its tokens using a crowdsale **ICO** (short for **Initial Coin Offering**).

We will build a smart contract for our ERC20 standard token using the OpenZeppelin libraries. Investors will be able to buy tokens from ICO smart contracts and, in return, they will receive newly minted tokens.

We will cover the following topics in this chapter:

- Building the features of our token from scratch
- Choosing the appropriate OpenZeppelin libraries to use
- Creating an ICO smart contract for our token
- Writing test cases for our smart contract
- Deploying a contract on testnet
- Getting investors to send ether to the contract to mint new tokens

Technical requirements

The following are the tools and libraries we are going to use in this chapter:

- You should have the latest version of node installed; version v8.11.3 or higher is preferred
- We use OpenZeppelin version 2.2.0 library contract files
- Truffle v5.0.4 or higher
- Ganache CLI v6.2.5 or higher
- MetaMask Google Chrome plugin version 6.3.1 or higher

In the Truffle project itself, we will need some npm packages to be installed, which will be required when writing test cases:

- Truffle HD wallet provider 1.0.6 or higher
- OpenZeppelin Test Helpers package 0.3.1 or higher
- Chai test framework 4.2.0 or higher
- Big Number package 2.0.0 or higher

The source code related to this chapter can be found on GitHub at https://github.com/ PacktPublishing/Mastering-Blockchain-Programming-with-Solidity/tree/master/ Chapter12.

Features of our token

To create a new token or any ICO contract, the first thing is to list down all the features of the contract that you require. For example, the features of an ICO contract can be pre-minted, mintable on-demand, burnable, capped crowdsale, and time-based crowdsale. All these features have been discussed in Chapter 9, *Deep Dive into the OpenZeppelin Library*. You can refer to that chapter for more details on these contracts.

Ideally, when you create a new contract or token, you have the specification documentation of the features and conditions to implement in the code. However, as we are going to learn how to create our own token, we need to jot down our own specifications. Now, let's build the specification of our token that we will create in this chapter:

- **Token contract specification**:
  - We want to build a token that should be fungible up to *18 decimal* places. Being a fungible token, we would use the ERC20 standard token.
  - The token name should be Mastering Solidity Token.
  - The **Mastering Solidity Token (MST)** token can be purchased initially from the ICO/crowdsale.
  - The token's symbol should be **MST**.
  - The MST token holders are allowed to burn their own tokens and the tokens for which they have received the approvals.
  - The deployer/owner of the token contract can pause and unpause the MST token transfers.

- **Crowdsale contract specification**:
  - The ICO should start on a specific date and last for 90 days from the start date.
  - The MST tokens can be bought during the crowdsale period only. Before or after this period, the crowdsale tokens cannot be purchased from the contract.
  - In the ICO, the rate of an MST token is such that, for 1 ETH, you will get 1,000 MST.
  - Once the crowdsale has finished, no more MST tokens should be minted.
  - The hard-cap of the token is limited to 10 million MST tokens. This means that at most, 10,000 ETH can be raised using this crowdsale.
  - The owner of the crowdsale will receive the ETH raised during the crowdsale.
  - The owner of the crowdsale can use or transfer the ETH raised during the crowdsale.

Now, as we know the features of the token and ICO crowdsale, we can start designing the contract by finding the right tools and libraries for the project.

Contract architecture design

Once you have the system specification written, the first thing is to find the right tools and libraries that should be used to build the project from scratch. It is always recommended that you look for existing libraries that are battle tested and easy to maintain in the project. The OpenZeppelin Solidity contract libraries provide many battle-tested contracts and libraries that are readily available to use in your project.

Our project requires ERC20 and crowdsale-related libraries, which are readily available in the OpenZeppelin library package. We have learned about these ERC20 and crowdsale-specific libraries in Chapter 9, *Deep Dive into the OpenZeppelin Library*.

There are many ERC20 and crowdsale-related libraries in the OpenZeppelin library package. Now, let's walk through how we can choose and use these libraries from the OpenZeppelin package to build our features.

We can divide our project requirements and specification into two modules. The first part of the project is to build MST token, and the second part is to raise funds using a crowdsale contract that mints MST tokens.

Designing an ERC20 MST token

As per the specification of the token, let's first find the libraries we will use to build our MST token:

- As per the specification, our MST token will be a fungible token, in which each MST token can be fungible up to 18 decimal places. For the fungible token, we should use the ERC20 token implementation provided by the OpenZeppelin library. We will use the ERC20.sol contract present in the OpenZeppelin library for our project.

- As we have seen in Chapter 9, *Deep Dive into the OpenZeppelin Library*, the chapter related to the ERC20 token standard, that is, ERC20.sol, does not provide a way to specify the token metadata such as token name, symbol, and decimals. To store this metadata, OpenZeppelin provides the ERC20Detailed.sol contract. We will use this contract to specify and initialize the token metadata; token name, token symbol, and the number of decimals each token is fungible up to.

- As per the specification, we require that the MST token holders should be allowed to burn their own tokens. For this feature, in OpenZeppelin, ERC20Burnable.sol is best suited. This contract allows a token holder to burn their own token and the tokens they have received approvals for.

- The owner/deployer of the MST token would need a feature, with which they can pause and unpause the MST token transfers. When the transfers are paused, no one is allowed to transfer their tokens. Once it is unpaused, every token holder is allowed to transfer or approve their MST tokens. Only the owner of the token or a person having PauserRole assigned is allowed to pause or unpause the transfers. The ERC20Pausable.sol contract should be used for this feature.

- In the crowdsale contract, whenever we receive the ether, the contract should mint new MST tokens, because MST tokens are minted on-demand during the crowdsale. The MST token is not a pre-minted token, in which tokens are minted before the crowdsale starts. For this feature, we would require our MST token to be a mintable token. We will use the ERC20Mintable.sol contract, so that the MinterRole can mint new MST tokens.

After ascertaining the best-suited library files, the following is a complete list of the contract files that our project requires for the MST token:

- ERC20.sol: For fungible MST tokens and the basic transfer and approval needs of the token.

- ERC20Detailed.sol: This allows MST tokens to have metadata and stores the token name, symbol and decimals.

- `ERC20Burnable.sol`: This allows MST token burning by the token holders.
- `ERC20Pausable.sol`: This allows an owner or `PauserRole` to pause and unpause the token transfers.
- `ERC20Mintable.sol`: This allows an owner or `MinterRole` to mint new MST tokens during the crowdsale.

The location for each of the contracts listed previously in the OpenZeppelin library is as follows:

Contract file name	Contract file location in openzeppelin-solidity library
ERC20.sol	openzeppelin-solidity/contracts/token/ERC20/ERC20.sol
ERC20Detailed.sol	openzeppelin-solidity/contracts/token/ERC20/ERC20Detailed.sol
ERC20Burnable.sol	openzeppelin-solidity/contracts/token/ERC20/ERC20Burnable.sol
ERC20Pausable.sol	openzeppelin-solidity/contracts/token/ERC20/ERC20Pausable.sol
ERC20Mintable.sol	openzeppelin-solidity/contracts/token/ERC20/ERC20Mintable.sol

We will use these contracts to build our MST token contract. However, there are indirect contracts, interfaces, and libraries that are also used while using these contracts. These indirectly used contracts are as follows:

- `SafeMath.sol`: The math library that is internally used by the ERC20 contract to prevent against integer overflow and underflow attacks
- `IERC20.sol`: The ERC20 interface, which defines the ERC20 standard functions and events
- `Roles.sol`: The general-purpose library to maintain roles
- `MinterRole.sol`: The contract to maintain `MinterRole`
- `PauserRole.sol`: The contract to maintain `PauserRole`
- `Pausable.sol`: The contract allows a `PauserRole` to pause and unpause

 As a general practice, a developer must know the directly used contracts and the indirectly used contracts, since the link between these contracts must be known to the developer to understand the behavior of the contracts.

Designing an MST crowdsale contract

As per the specification of the crowdsale, let's first find the libraries we will use to build our MST crowdsale:

- As per the specification, we would want to mint new MST tokens on-demand when someone sends ether to the crowdsale contract. For this, we would need the `MintedCrowdsale.sol` contract from the OpenZeppelin library. The only prerequisite for this contract is that it requires the `MinterRole` on the MST token contract. At the time of deployment of this contract, we will assign `MinterRole` to the contract. However, the `MintedCrowdsale.sol` contract does not impose any cap on minting new tokens. You can mint any number of tokens using this contract.

- To apply the cap on minting new tokens, we would need the `CappedCrowdsale.sol` contract. This contract takes the cap as a parameter to the constructor and will not allow the minting of new MST tokens once the cap is reached.

- As per the specification, we would want to open the crowdsale for a duration of 90 days. Before and after that duration, the minting of new MST tokens should not be allowed. For this requirement, we would use the `TimedCrowdsale.sol` contract. This contract takes the start and closing time of the crowdsale. Once the contract is deployed, it will automatically start accepting the ether upon the start time of the crowdsale and stop accepting any further ether to the contract once the closing time is reached.

After finding out the best-suited library files, the following is the complete list of contract files our project requires for the MST crowdsale:

- `MintedCrowdsale.sol`: To mint new MST tokens when ether is received at the crowdsale contract
- `CappedCrowdsale.sol`: To allow the minting of new MST tokens less than or equal to the cap set
- `TimedCrowdsale.sol`: To allow the minting of new MST tokens only between the opening and closing time duration of the crowdsale

The location for each of the contracts listed here in the OpenZeppelin library is as follows:

Contract file name	Contract file location in openzeppelin-solidity library
MintedCrowdsale.sol	openzeppelin-solidity/contracts/crowdsale/emission/MintedCrowdsale.sol
CappedCrowdsale.sol	openzeppelin-solidity/contracts/crowdsale/validation/CappedCrowdsale.sol
TimedCrowdsale.sol	openzeppelin-solidity/contracts/crowdsale/validation/TimedCrowdsale.sol

We will use these contracts to build our MST crowdsale contract. However, there are indirect contracts, interfaces and libraries that are also used while using these contracts. These indirectly used contracts are as follows:

- `SafeMath.sol`: The math library that is internally used by the `Crowdsale` contract to present against integer overflow and underflow attacks.
- `Crowdsale.sol`: The basic implementation of the crowdsale contract.
- `ReentrancyGuard.sol`: To prevent the reentrancy attacks, used in `Crowdsale` contract.
- `IERC20.sol`: Interface to connect to the token contract, used in `Crowdsale` contract.

We have now designed the MST token and MST crowdsale architecture and found which libraries to use for which purpose. Now, we can start setting up the Truffle project.

Setting up the Truffle project

We learned about Truffle in `Chapter 5`, *Using Ganache and the Truffle Framework*. To set up a new Truffle project, we have to do the following:

1. Execute the following command in an empty folder:

   $ truffle init

   This will initialize the Truffle project-related files and folders under the current directory.

2. Initialize the `npm` packages in the Truffle project. To enable the project to install `npm` packages, you will need to execute the following command:

   $ npm init

   The command will ask for certain details related to your package that you can provide, and it will create a `package.json` file under the current folder.

3. Next, we need to install the dependencies of the project that we are going to use. As we have designed the contract's architecture, looking at that, we would need libraries from OpenZeppelin.

To install the latest version of the OpenZeppelin libraries, execute the following command:

```
$ npm install openzeppelin-solidity
```

The preceding command will install the OpenZeppelin package under the `node_modules` folder. You can check the version of the OpenZeppelin package installed in the `package.json` file. You can also change the version to use in this file.

By setting up the Truffle project, it has created a `truffle-config.js` file. We would need to check and update the configuration according to our needs. Now, let's update the configuration.

 It is recommended that the latest version of OpenZeppelin libraries should be used. Also, developers should keep checking the recent issues fixed in the latest versions of OpenZeppelin.

Updating configuration

In the `truffle-config.js` file, you will find the compilers' settings. You will need to uncomment the required parameters for this setting. As we are going to use Solidity compiler version `^0.5.0`, we have updated the configuration accordingly. We also want our deployed contracts to be optimized, hence, we have enabled that flag. The following is the configuration for compilers in the `truffle-config.js` file:

```
compilers: {
  solc: {
    version: "^0.5.0",
    docker: false,
    settings: {
     optimizer: {
       enabled: true,
       runs: 200
     },
     evmVersion: "byzantium"
    }
  }
}
```

Also, uncomment the `development` network configuration in the `truffle-config.js` file. After this configuration setup, you are ready to write your contract in the `contracts` folder present in the Truffle project. Now, let's create our MST Token and MST crowdsale contracts.

Creating Solidity contracts

To create the contracts for MST token and MST crowdsale, you need to create `MSTToken.sol` and `MSTCrowdsale.sol` files under the `contracts` folder. According to the contract feature, you can have any appropriate name for your contracts.

You can use Remix IDE or any other file editor to write the Solidity files. However, some plugins are supported for certain editors, for example, IntelliJ IDEA plugin, Visual Studio Plugin, and Package for Sublime Text. Now, let's create the `MSTToken.sol` contract.

The MSTToken.sol contract

According to our contract design and architecture, we created the `MSTToken.sol` contract. The following is the code for the `MSTToken.sol` contract:

```solidity
pragma solidity >=0.5.0 <0.6.0;

import "openzeppelin-solidity/contracts/token/ERC20/ERC20Pausable.sol";
import "openzeppelin-solidity/contracts/token/ERC20/ERC20Detailed.sol";
import "openzeppelin-solidity/contracts/token/ERC20/ERC20Burnable.sol";
import "openzeppelin-solidity/contracts/token/ERC20/ERC20Mintable.sol";

contract MSTToken is
    ERC20Burnable, ERC20Pausable, ERC20Mintable, ERC20Detailed {

    constructor(
        string memory _name,
        string memory _symbol,
        uint8 _decimals
    )
        ERC20Detailed(_name, _symbol, _decimals)
        public
    {
        // solium-disable-previous-line no-empty-blocks
    }
}
```

At the first line of the contract file, we defined the Solidity compiler version to be used. For this contract, we require a compiler version greater than or equal to version 0.5.0, and lesser than version 0.6.0. Next, we imported the OpenZeppelin library contracts that we are going to use in this contract.

At the start of the contract block, we defined the contract name, `MSTToken`. This contract inherits from the `ERC20Burnable`, `ERC20Pausable`, `ERC20Mintable`, and `ERC20Detailed` contracts. We discussed the features of these individual contracts in detail in this chapter's *Contract architecture design* section.

The constructor of the contract takes three parameters—the name (for contract name), symbol (for contract symbol), and decimals (number of decimals to support). These parameters are passed to the constructor of the `ERC20Detailed` contract. This will initialize the MST token metadata at the time of deployment. The constructor block is empty and, to remove the `solium` warning message, we added the `solium` disable comment.

Once the contract is deployed by the deployer's wallet, the deployer will be, by default, assigned `PauserRole` and `MinterRole`. These roles will allow the deployer to pause and unpause the transfers, and will also allow them to mint any number of tokens.

Our MST token contract has been created. Now, we will create the `MSTCrowdsale.sol` contract.

The MSTCrowdsale.sol contract

According to our contract design and architecture, we will create the `MSTCrowdsale.sol` contract. The following is the code for the `MSTCrowdsale.sol` contract:

```
pragma solidity >=0.5.0 <0.6.0;

import "openzeppelin-
solidity/contracts/crowdsale/validation/TimedCrowdsale.sol";
import "openzeppelin-
solidity/contracts/crowdsale/emission/MintedCrowdsale.sol";
import "openzeppelin-
solidity/contracts/crowdsale/validation/CappedCrowdsale.sol";
import "openzeppelin-solidity/contracts/token/ERC20/IERC20.sol";

contract MSTCrowdsale is
    CappedCrowdsale, TimedCrowdsale, MintedCrowdsale {
```

```
constructor(
    uint256 _rate,
    address payable _wallet,
    IERC20 _token,
    uint256 _openingTime,
    uint256 _closingTime,
    uint256 _cap
)
    Crowdsale(_rate, _wallet, _token)
    TimedCrowdsale(_openingTime, _closingTime)
    CappedCrowdsale(_cap)
    public
{
    // solium-disable-previous-line no-empty-blocks
}
}
```

At the first line of the contract file, we defined the solidity compiler version to be used. For this contract, we require a compiler version greater than or equal to version 0.5.0, and lesser than version 0.6.0. Next, we imported the OpenZeppelin library contracts that we are going to use in this contract.

At the start of the contract block, we defined the contract name, MSTCrowdsale. This contract inherits from the CappedCrowdsale, TimedCrowdsale, and MintedCrowdsale contracts. We discussed the features of these individual contracts in detail in this chapter's *Contract architecture design* section.

In the constructor of the contract, the following arguments are passed:

- _rate: The rate per 1 wei ETH to mint new MST tokens
- _wallet: The wallet address to which the received ETH at the crowdsale contract will be forwarded
- _token: The MST token contract address
- _openingTime: The opening time of the crowdsale
- _closingTime: The closing time of the crowdsale
- _cap: The max cap of ETH received in the contract

Once the MSTCrowdsale contract is deployed, it must have the MinterRole assigned for the MSTToken contract. Using this role, MSTCrowdsale will be able to mint new MST tokens for the investors. We will have these deployment-specific configurations and roles in the migration scripts, which we will see in further sections of this chapter.

The Solidity contract creation is complete. Now, let's compile the project.

Compiling contracts

To compile the project, execute the following command in the Truffle project:

```
$ truffle compile
```

As you can see from the following the output of the `truffle compile` command, it compiles all the contracts, including the OpenZeppelin contracts that we are using in the project:

```
✔   truffle compile

Compiling your contracts...
===========================
> Compiling ./contracts/MSTCrowdsale.sol
> Compiling ./contracts/MSTToken.sol
> Compiling ./contracts/Migrations.sol
> Compiling openzeppelin-solidity/contracts/access/Roles.sol
> Compiling openzeppelin-solidity/contracts/access/roles/MinterRole.sol
> Compiling openzeppelin-solidity/contracts/access/roles/PauserRole.sol
> Compiling openzeppelin-solidity/contracts/crowdsale/Crowdsale.sol
> Compiling openzeppelin-solidity/contracts/crowdsale/emission/MintedCrowdsale.sol
> Compiling openzeppelin-solidity/contracts/crowdsale/validation/CappedCrowdsale.sol
> Compiling openzeppelin-solidity/contracts/crowdsale/validation/TimedCrowdsale.sol
> Compiling openzeppelin-solidity/contracts/lifecycle/Pausable.sol
> Compiling openzeppelin-solidity/contracts/math/SafeMath.sol
> Compiling openzeppelin-solidity/contracts/token/ERC20/ERC20.sol
> Compiling openzeppelin-solidity/contracts/token/ERC20/ERC20Burnable.sol
> Compiling openzeppelin-solidity/contracts/token/ERC20/ERC20Detailed.sol
> Compiling openzeppelin-solidity/contracts/token/ERC20/ERC20Mintable.sol
> Compiling openzeppelin-solidity/contracts/token/ERC20/ERC20Pausable.sol
> Compiling openzeppelin-solidity/contracts/token/ERC20/IERC20.sol
> Compiling openzeppelin-solidity/contracts/token/ERC20/SafeERC20.sol
> Compiling openzeppelin-solidity/contracts/utils/Address.sol
> Compiling openzeppelin-solidity/contracts/utils/ReentrancyGuard.sol

    > compilation warnings encountered:

openzeppelin-solidity/contracts/crowdsale/Crowdsale.sol:138:5: Warning: Function state mutability can be restricted to pure
    function _preValidatePurchase(address beneficiary, uint256 weiAmount) internal view {
    ^ (Relevant source part starts here and spans across multiple lines).

> Artifacts written to /Users/jitendra/Documents/MasteringSolidity/Mastering-Solidity/Chapter12/build/contracts
> Compiled successfully using:
    - solc: 0.5.7+commit.6da8b019.Emscripten.clang
```

The output of the Truffle compile command

As you can see in the preceding screenshot, the `truffle compile` command returned compilation warnings. This warning is present in the `Crowdsale.sol` contract of the OpenZeppelin library. As per the warning, it is suggesting the use of the `pure` keyword for the `_preValidatePurchase()` function. However, the function can be overridden in the implementing contracts, hence it cannot be defined as `pure`. You can ignore this warning message.

Our contracts are ready and compiled. Now, we can write the migration scripts, which will be used in the deployment of the contracts.

Writing a migration script

After writing the Solidity contracts, we should write the migration scripts. The migration scripts define the contract deployment setup. The migration script for the MSTToken and MSTCrowdsale contract is created in the migrations folder. The migration script is named 2_deployCrowdsale.js, and you can find this file under the migrations folder. We did discuss the migration scripts in Chapter 5, *Using Ganache and the Truffle Framework, Writing contract migration scripts* section.

The following is the code of the 2_deployCrowdsale.js migration script:

```
const MSTToken = artifacts.require("MSTToken");
const MSTCrowdsale = artifacts.require("MSTCrowdsale");

var BigNumber = require('big-number');

module.exports = async function(deployer, network, accounts) {
  var owner = accounts[0];
  var wallet = accounts[9];

  //1. Deploy MSTToken
  await deployer.deploy(MSTToken, "Mastering Solidity Token", "MST", 18);

  //2. Deploy MSTCrowdsale
  var milliseconds = (new Date).getTime(); // Today time
  var currentTimeInSeconds = parseInt(milliseconds / 1000);
  var oneDayInSeconds = 86400;
  var openingTime = currentTimeInSeconds + 60; // openingTime after
  //a minute
  var closingTime = openingTime + (oneDayInSeconds * 90); // closingTime
  //after 90 days
  var rate = 1000; //1000 MST tokens per ether
  var cap = BigNumber(10000).pow(18); // 10000 ** 18 = 10000 ether

  await deployer.deploy(
    MSTCrowdsale,
    rate,
    wallet,
    MSTToken.address,
    openingTime,
    closingTime,
    cap
  );

  //3. Owner Adds MinterRole for MSTCrowdsale
  var mstToken = await MSTToken.deployed();
```

```
mstToken.addMinter(MSTCrowdsale.address);

//4. Owner Renounce Minter
mstToken.renounceMinter();

};
```

As you can see in the preceding migration script, we are using the `BigNumber` package. To install this package, you would need to execute the following command under your Truffle project:

```
$ npm install big-number
```

The preceding command will install the required package dependencies in the project.

Next, in the migration script, you can see the following sequence of events:

1. The script deploys the `MSTToken` contract using the `owner` wallet (by default, the `accounts[0]` wallet). We provide the token metadata details for the `MSTToken` constructor.

2. The script deploys the `MSTCrowdsale` contract and passes the parameters to the constructor of the contract.

3. The `owner` account (`account[0]`) has the `MinterRole` assigned by default as that account deployed the `MSTToken` contract. An account having `MinterRole` can assign any other account as a `MinterRole`. In this step, the owner has the `MinterRole` for `MSTToken`. He is adding a new `MinterRole` to the `MSTCrowdsale` contract. So that the `MSTCrowdsale` contract can request the `MSTToken` contract to mint new MST tokens.

4. By doing the third step, there are two accounts that have the `MinterRole`— the owner account and the `MSTCrowdsale` contract. As we want only `MSTCrowdsale` to have the `MinterRole`, the owner should renounce his `MinterRole`. After this `renounceMinter()` function call by the account owner, there will only be a single `MinterRole` assigned to `MSTCrowdsale`. And `MSTCrowdsale` has no way of giving `MinterRole` to any other account, hence, the setup is secure and new MST tokens can only be minted via `MSTCrowdsale`. There is no other way possible to mint MST tokens.

Now, let's execute the migration script and deploy the contracts.

Running Ganache

To execute the migration script in the development environment, the local blockchain should be running. To run the Ganache CLI tool, execute the following command:

```
$ ganache-cli
```

This will start the local blockchain at port 8545. You need to keep this running in the command line.

Now, let's run the migration scripts.

Running migration scripts

To execute the migration scripts present in the `migrations` folder, you will need to execute the following command:

```
$ truffle migrate
```

The preceding command will start the migration process and begin executing migration scripts. As the command does not have any network parameters mentioned, it will execute migration using the `development` network. The configuration of the `development` network is present in `truffle-config.js` by default, ensure that it is uncommented.

The execution of the `migrate` command will generate the following output:

```
Starting migrations...
======================
> Network name:    'development'
> Network id:      1554358334334
> Block gas limit: 0x6691b7


1_initial_migration.js
======================
Deploying 'Migrations'
----------------------
 > transaction hash:
0x88b175f87c78066f061221a34d621ffc3ec01a805fb8599b585514bb84e9cbb3
 > Blocks: 0 Seconds: 0
 > contract address:    0xF5815cd93B1105EB2A65682aFB9019F78Cb99f9b
 > account:             0x198e6dD7c8947fF22debe2c78378Bef04F2E64ee
 > balance:             99.99557658
 > gas used:            221171
 > gas price:           20 gwei
 > value sent:          0 ETH
 > total cost:          0.00442342 ETH
```

```
> Saving migration to chain.
> Saving artifacts
------------------------------------
> Total cost: 0.00442342 ETH
```

2_deployCrowdsale.js
====================
Deploying 'MSTToken'

```
> transaction hash:
0x47cad7c5f516882f5dee7cdc2a6c7b55df621991b420b1676298f56e1e09e5e5
> Blocks: 0 Seconds: 0
> contract address: 0xDc0E5080856ffA2357ea0C91d14C0eF7989582Fd
> account: 0x198e6dD7c8947fF22debe2c78378Bef04F2E64ee
> balance: 99.97106936
> gas used: 1183432
> gas price: 20 gwei
> value sent: 0 ETH
> total cost: 0.02366864 ETH
```

Deploying 'MSTCrowdsale'

```
> transaction hash:
0x8e7f854877df56a15265981f89c25b008c396282e3fe71afab3e97baadde3151
> Blocks: 0 Seconds: 0
> contract address: 0x1871770d6C483cCfeCda798A493a73fAFdACfDd7
> account: 0x198e6dD7c8947fF22debe2c78378Bef04F2E64ee
> balance: 99.95986886
> gas used: 560025
> gas price: 20 gwei
> value sent: 0 ETH
> total cost: 0.0112005 ETH
> Saving migration to chain.
> Saving artifacts
------------------------------------
> Total cost: 0.03486914 ETH
```

Summary
=======
```
> Total deployments: 3
> Final cost: 0.03929256 ETH
```

As you can see in the output, first it shows metadata related to Network name, Network id, and Block gas limit, which will be used while executing the migration script. After that, it starts executing migration scripts one by one according to the number used in the filename, starting from 1_ and so on. First, it executes the 1_initial_migration.js migration file, which deploys the Migration contract. The deployment transaction detail is shown. Next, it executes the 2_deployCrowdsale.js migration file, which deploys the MSTToken and MSTCrowdsale contracts in order. Details for these transactions are also included in the output. At last, you can see in the summary that the migration scripts deployed three contracts in total and that it took 0.03929256 ETH to deploy the contracts.

Now, let's move on to creating the test cases for the project.

Writing test cases

To write the test cases for our contracts, we will need to install some dependencies. We will use the chai test framework to write tests in JavaScript.

1. To install the chai test framework, execute the following command:

    ```
    $ npm install chai
    ```

2. Next, we install the openzeppelin-test-helpers package. This package is built by OpenZeppelin and is used in writing the test cases. The package has some helper utilities that will be required while writing test cases for our project.

    To install the OpenZeppelin Test Helpers package, execute the following command:

    ```
    $ npm install openzeppelin-test-helpers
    ```

In this project, we wrote two test files for our contracts. These files are present under the project's test folder:

- MSTToken.test.js: This file contains test cases for the MSTToken contract.
- MSTCrowdsale.test.js: This file contains test cases for the MSTCrowdsale contract.

The preceding file contents are sufficiently large to accommodate in the book. You can refer to these files at their GitHub location: https://github.com/PacktPublishing/Mastering-Blockchain-Programming-with-Solidity/tree/master/Chapter12/test.

Once the test cases are written, you should run these test cases and ensure that everything is working as expected. To run the tests on the development network, execute the following command:

```
$ truffle test
```

The following is the output of the test cases:

```
Contract: MSTCrowdsale
  as TimedCrowdsale
    ✓ should have crowdsale openTime
    ✓ should have crowdsale closingTime
    ✓ should not hasClosed
    ✓ should not isOpen
    ✓ should Open crowdsale (72ms)

Contract: MSTCrowdsale
  as Crowdsale
    ✓ should default to zero
    ✓ should open crowdsale (65ms)
    ✓ should mint MST when ETH received (118ms)

Contract: MSTToken
  as ERC20Detailed
    ✓ should have token name
    ✓ should have token symbol
    ✓ should have token decimals
  as ERC20Mintable
    ✓ owner should have minter role
    ✓ minter should mint (70ms)
  as ERC20
    ✓ should return totalSupply
    ✓ should return balanceOf (67ms)
    ✓ should transfer (110ms)
    ✓ should fail transfer (119ms)
    ✓ should approve (94ms)
    ✓ should return allowance (110ms)
    ✓ should transferFrom (346ms)
    ✓ should increaseAllowance (282ms)
    ✓ should decreaseAllowance (371ms)
  as ERC20Pausable
    ✓ should be unpaused by default
    ✓ should pause (78ms)
    ✓ should unpause (144ms)
    ✓ should not allow transfer when paused (166ms)
    ✓ should not allow transferFrom when paused (219ms)
    ✓ should not allow approve when paused (155ms)
    ✓ should not allow increaseAllowance when paused (150ms)
    ✓ should not allow decreaseAllowance when paused (176ms)
  as ERC20Burnable
    ✓ should burn (89ms)
    ✓ should burnFrom (223ms)


32 passing (7s)
```

The output of the Truffle test command

As you can see, there are 32 test cases present and all the tests passed. Once the tests are working and the required parts of the project are covered with the test cases, we can consider the project development complete.

However, now you can run the other tools on your project, improve the quality further, and fix any issues present in the project.

Deploying contracts on testnet

As of now, we have built the project using Truffle and a local instance of Ganache for local blockchain. However, Ethereum does provide public test networks (also known as testnets) for developers to test their contract for a production-like environment. To use these testnets, you need to connect your Truffle project to one of these testnets. The names of these testnets are Ropsten, Kovan, Rinkeby, and Goerli.

There are two ways to use one of these testnets:

- You can download and configure your local Ethereum wallet client software and connect to one of the testnets. After configuring the client software, it will download the whole blockchain ledger and keep syncing the blockchain on your machine.
- Another way is to use the Infura account and connect to one of these testnets. When you use Infura links, you do not need to download the blockchain ledger on your machine. This is the easiest and most user-friendly way to use testnets as well as mainnet.

To use Infura you will need to sign up on `infura.io` and use the different Infura links provided. Now, let's learn about Infura APIs.

Infura APIs

We did discuss Infura in detail in the *Using Infura* section in `Chapter 5`, *Using Ganache and the Truffle Framework*. Once you have registered on Infura, you need to create a new project. It is recommended that for each of your separate DApp projects, you should create a different project on Infura and use the links generated for that project.

The following are the Infura links provided by Infura on their website, once you create a new project:

```
KEYS


  PROJECT ID                                   PROJECT SECRET  ⓘ

  138bdb3fed8f4c96b812327d76be8a08  ▢          f67b16e6b06e4da8bf63ddd1d756906c  ▢


  ENDPOINT    RINKEBY ⌄

  rinkeby.infura.io/v3/138bdb3fed8f4c96b812327d76be8a08  ▢
```

The keys generated by infura.io

Once you have the link, you can use that link in your Truffle configuration file. Let's modify the Truffle project configuration file to use this link.

Updating the configuration

You can uncomment the appropriate sections present in the default Truffle configuration file. By default, the Ropsten testnet network configuration is present in the Truffle configuration file. Similarly, we have prepared the configuration for the Rinkeby testnet network:

```
const HDWalletProvider = require('truffle-hdwallet-provider');
const infuraKey = "138bdb3fed8f4c96b812327d76be8a08";

const fs = require('fs');
const mnemonic = fs.readFileSync(".secret").toString().trim();
...
...
rinkeby: {
  provider: () => new HDWalletProvider(mnemonic,
  'https://rinkeby.infura.io/v3/${infuraKey}', 0, 20),
  network_id: 4,
  gas: 6000000,
  gasPrice: 10000000000,
  confirmations: 2,
  timeoutBlocks: 200,
  skipDryRun: true
},
```

Let's understand each configuration in detail:

- `rinkeby`: This is the network configuration name that will be used in the `truffle` command to specify the network name. The rest of the configuration for this network is under here.
- `provider`: As you can see, for a wallet provider, we are using `HDWalletProvider`. You need to install this provider; you can follow the *Installing dependencies* section described after this section to install this dependency. There are some benefits associated with using `HDWalletProvider`. When your wallets are generated using BIP39 mnemonic keywords, you can use this provider. The provider also takes the wallet mnemonic as the first argument. Secondly, it takes the Infura URL provided by Infura along with the key given by Infura. The third and fourth parameters are set to 0 and 20, meaning that you want to load 20 EOA accounts from index 0 using the given mnemonic.
- `network_id`: This is the network ID of the test network to be used. Rinkeby has the network ID 4.
- `gas`: This is the gas limit in gas units to be used per transaction sent from `truffle`. The gas limit is set to 6 million.
- `gasPrice`: This is the gas price in wei to be used per transaction sent from `truffle`. The gas price is set to 10 giga wei.
- `confirmations`: This is the number of confirmations to wait after the transaction is included in the block. The block confirmation is set to 2.
- `timeoutBlocks`: This is the timeout of the script in the number of blocks once it has started. This is set to 200 in the configuration.
- `skipDryRun`: Should the dry run of the command execution be skipped? In the preceding configuration, we have this set to `true`, meaning that we do not want to perform the dry run of the command execution.

We have used mnemonics in the preceding configuration. Now, let's learn where to keep the mnemonics for the `truffle` project.

Setting up wallet mnemonics

The `truffle` configuration file reads the mnemonics from a `.secret` file present in the `truffle` project directory. This is done using the following lines already present in the `truffle-config.js` file. These lines could be commented initially. However, you can uncomment those lines according to your needs:

```
const fs = require('fs');
const mnemonic = fs.readFileSync(".secret").toString().trim();
```

The preceding script reads the `.secret` file from the filesystem and sets in the mnemonic variable. This variable can be used in the configuration to read the BIP39 mnemonic keyword.

If this file is not present, you should create this file under the `truffle` project and incorporate your BIP39 mnemonic seed keywords in plain text format. In this project, we have created a `.secret` file; the content of that file is as follows:

```
quality course comfort canyon bunker chief boring afraid original crop
north series
```

Using the preceding mnemonic seed keywords, wallets will be generated and used in the `truffle` project. You should not use the same seed words shared previously. Generate your own mnemonic by setting up a new wallet on MetaMask.

Installing dependencies

Next, we install the `truffle-hdwallet-provider` package. This package is needed to have support for the **Hierarchical Deterministic (HD)** wallets. By default, using the `development` network in Truffle, it uses the wallets initialized in the Ganache GUI and the Ganache CLI. However, when you want your contracts to be deployed on a testnet (Kovan, Rinkeby, Ropsten) or on mainnet, you will need support for HD wallets. Using this package, you can build wallets using mnemonic keywords. We did discuss the secret seed phrase mnemonics in Chapter 4, *Learning MetaMask and Remix, Installing and setting up MetaMask plugins* section.

You can install this package by executing the following command:

```
$ npm install truffle-hdwallet-provider
```

The preceding command will install the required files for `HDWalletProvider` under the `node_modules` folder present under the project folder.

Let's now deploy these contracts on the Rinkeby testnet.

Deploying contracts on testnet

We have everything set up for the Rinkeby testnet. Now, we can deploy the contracts on the Rinkeby test network by executing the following command:

```
$ truffle migrate --network rinkeby
```

The command execution will take some time to process, as the contracts are deployed on the Rinkeby test network. The following is the command output you will see during the contract deployment. Note that the transaction hash and contract addresses will be different when you execute the command:

```
Compiling your contracts...
===========================
> Everything is up to date, there is nothing to compile.

Starting migrations...
======================
> Network name:    'rinkeby'
> Network id:      4
> Block gas limit: 0x6acfc0

1_initial_migration.js
======================

  Deploying 'Migrations'
  ----------------------
  > transaction hash:
0x85fc5089f54ae26ab73198f0550ef24beeb0229b3beea91640f856109ffe1f5a
    > Blocks: 1 Seconds: 13
    > contract address: 0x8f986A1EB6185b0039ccA6d6686c9Df7d8794B10
    > block number: 4489054
    > block timestamp: 1559465484
    > account: 0x33978ED1D8c28112f1a4f57F5a3d64d5ff3a802F
    > balance: 9.991932538197947632
    > gas used: 223519
    > gas price: 10 gwei
    > value sent: 0 ETH
    > total cost: 0.00223519 ETH
```

```
Pausing for 2 confirmations...
_____
> confirmation number: 1 (block: 4489055)
> confirmation number: 2 (block: 4489056)

Command output truncated...
```

The preceding log is truncated as it was big enough. You can find the full command output on GitHub at `https://github.com/PacktPublishing/Mastering-Blockchain-Programming-with-Solidity/blob/master/Chapter12/output/truffle_migrate.txt`.

When we executed the migration scripts against the local Ganache instance, we got a similar output. Here you can see that the execution order is the same. However, this also shows the confirmation of the testnet blockchain. The remainder of the process is the same.

Once your contract migration script is executed, you can check the network status by executing the following command:

```
$ truffle networks
```

The output of the preceding command is as follows:

```
Network: rinkeby (id: 4)
  MSTCrowdsale: 0xd9F5E569d8D24C5473CcC7B7061f3795b1347132
  MSTToken: 0xd4d81EB4e1C443450cb91E920aB5cb7d6F795bFF
  Migrations: 0x8f986A1EB6185b0039ccA6d6686c9Df7d8794B10
```

As you can see, under the `rinkeby` network name, it displays the names and addresses of the contracts deployed on testnet. Hence, your contracts are deployed on the test network and they are ready to be used.

We have covered the contract migration on test networks. Similarly, you can also use mainnet to deploy your contracts. To do this, you just need to update your Infura link to point it to mainnet.

Note that, just like testnet, on mainnet, every transaction also requires ether in your wallet. However, the mainnet ether has economic value.

Summary

In this chapter, we learned how to create your own ERC20 token from scratch. To do this we started with the specification of the token to list down the features of the token. Then, we moved on to choosing the appropriate contract library files that would help to reduce the development cycle of the contracts. We then developed the contracts in Solidity. After the development of the contracts, we created migration scripts and test cases for the project. We also deployed and tested the project on the local instance of the Ganache blockchain and on the Rinkeby testnet.

We followed every step required to build the project from scratch. To further enhance the development process, you should also use linters, such as Eslint and Solhint, to maintain the project's quality and remove unwanted issues from your project.

In the next chapter, we will learn about the different Solidity design patterns available. These patterns can be used by developers to eliminate the known issues from the contract.

Questions

1. Which version of the Truffle and OpenZeppelin libraries should be used?
2. Is keeping the secret key mnemonic in the `.secret` file secure?
3. Should we run test cases on testnets and mainnet?
4. Can I remove the build folder that is created during the compilation?
5. Can I run migration and tests without ether?
6. Which testnet should be used for testing?

Section 4: Design Patterns and Best Practices

4

In this section, the reader will understand advanced design patterns, general best practices, and security best practices.

The following chapters will be covered in this section:

- Chapter 13, *Solidity Design Patterns*
- Chapter 14, *Tips, Tricks, and Security Best Practices*

13
Solidity Design Patterns

The Solidity language is a contract-oriented programming language. There are many constructs required for contract interaction. Also, due to the limitations of the language, few data structure-specific design patterns are used.

In this chapter, you will learn about the different design patterns, that are grouped into different categories based on their usage. These categories are as follows:

- Security design patterns
- Creational patterns
- Behavioral patterns
- Economic patterns
- Life cycle patterns

Out of these categories, some of the design patterns that are mostly used in Solidity contracts are listed here:

- The withdrawal pattern to withdraw ether from the contract
- The factory contract pattern to create new contracts
- The state machine pattern to transition a contract from different states
- The tight variable packing to reduce gas consumption when using structs

Security patterns

There are a few security patterns that ensure the security of funds in a contract. Using these patterns, you can improve the security of the contract. Using these patterns, you could eradicate common security issues in the contract. In the year 2018, there were more than $1 billion smart contract-related hacks. The patterns that we are discussing are built to avoid developers mistakes while developing smart contracts.

Let's look at some of the security design patterns.

Withdrawal pattern

The withdrawal pattern is also known as a **pull-over-push pattern**. In this pattern, ether or token transfer (push) from the contract is avoided; rather, the user is allowed to pull ether or token from the contract.

There can be many contracts in which you want to send ether or token to multiple addresses or to a group of addresses. Sending ether from a contract via iterative or non-iterative methods is always going to cause problems, so this should be avoided. Consider the following `DividendContract` sample code of the contract. Using this contract, you can distribute the dividend to your investors:

```
contract DividendContract is Ownable {

    address[] public investors;

    function registerInvestor(address _investor) public onlyOwner {
        require(_investor != address(0));
        investors.push(_investor);
    }

    //Bad Practice
    function distributeDividend() public onlyOwner {
        for(uint i = 0; i < investors.length; i++) {
            uint amount = calculateDividend(investors[i]);
            investors[i].transfer(amount); //Push ether to user
        }
    }

    function calculateDividend(address _investor) internal returns(uint) {
        //Dividend calculation here
    }
}
```

The contract maintains the list of investors in the investors array. Only the owner of the contract can add investors who are eligible to receive the dividend. The ether present in the contract will be distributed as dividends. As you can see in the distributeDividend() function, there is an unbounded loop and it will iterate the number of times equals the investors array length. There could be multiple issues in the preceding code, such as the following:

- In the future, there could be an out-of-gas exception due to this loop as the amount of gas consumption would increase linearly according to the increase in the investor list, hence it's dangerous to do this iteratively. There is no way to remove an investor from an array; hence, once the loop starts consuming more gas units than a block gas limit, the transaction to this function call will always fail. It is dangerous to write loops like this way. This will lead to the locking of ether in the contract, assuming that there is no other way to take funds out of the contract.

- It is possible that the address of an investor is a contract address, which has a continually failing fallback function. This leads to whole transaction failure for the distributeDividend() function call each time. This also leads to the locking of ether in the contract.

To avoid the issues described in the preceding text, the contract should be designed in a way that a user should be able to claim their dividend from the contract. This way, it's the onus of the user to pull the funds. As you can see in the following code, we have the claimDividend() function, that can be called by anyone. However, only the user who has valid balances present in the contract can claim dividends, as shown in the following:

```
//Good Practice - Pull ether
function claimDividend() public {
    uint amount = balances[msg.sender];
    require(amount > 0);
    //Ensure to update balance before transfer
    //to avoid reentrancy attack
    balances[msg.sender] = 0;
    msg.sender.transfer(amount);
}
```

The contract should maintain or update the balances of each user or investor accordingly, so that they can withdraw the balance amount from the contract, as, in the preceding code, the balances mapping must be updated via other functions.

Let's take a look at when the withdrawal pattern should be applied.

Applicability

We should use the withdrawal pattern or pull-over-push pattern in the following cases, when:

- You want to send ether/token to multiple addresses from the contract.
- You want to avoid the risk associated with transferring ether/token from the contract to the users.
- You want to avoid paying transaction fees as we know, the transaction initiator has to pay the transaction fee to get their transaction included in the blockchain. Hence, you may want to avoid paying transaction fees for transferring ether or token (push transaction) from the contract to your users. Instead, you want your users to pay transaction fees (pull transaction) and get their share of ether/token withdrawn from the contract.

Access restriction pattern

As the name suggests, the access restriction design pattern restricts access to the functions of the contract based on roles. The Ethereum blockchain is public and all the addresses and transaction data is public to everyone. However, we can define the contract state variables as `private`; doing this it only restricts reading the state variable from the contract. Still, anyone can follow the transactions and can find the values of the `private` state variables present in the contract. To allow anyone to call a function present in the contract, it should be defined as `public` or `external`. However, when you need restricted access to a function, you should use modifiers to check for the access rights.

In the following example code, we have two `address` state variables—`owner` and `admin`. The access restriction requirements are as follows:

- The `owner` variable stores the address of the contract owner. Only the `owner` of the contract is allowed to change the `admin` address using the `changeAdmin()` function.
- The `admin` variables store the address of the contract admin.
- Both `owner` and `admin` are authorized to pause the contract via the `pause()` function:

Let's have a look at the AccessControl contract:

```
contract AccessControl {
    address public owner;
    address public admin;
    bool public paused;
    modifier whenPaused() {
        require(paused);
        _;
    }

    modifier whenNotPaused() {
        require(!paused);
        _;
    }

    modifier onlyOwner() {
        require(msg.sender == owner);
        _;
    }

    modifier onlyAdmin() {
        require(msg.sender == admin);
        _;
    }

    modifier onlyAuthorized() {
        require(msg.sender == owner || msg.sender == admin);
        _;
    }

    function changeAdmin(address _newAdmin) public onlyOwner {
        require(_newAdmin != address(0));
        admin = _newAdmin;
    }

    function pause() public onlyAuthorized whenNotPaused {
        paused = true;
    }
}
```

As you can see in the preceding code, the following is true:

- The changeAdmin() function is only allowed to be executed from the owner address as it is restricted with an onlyOwner modifier. This function is always allowed to be called from owner, as it is not protected with either the whenPaused or whenNotPaused modifiers.
- The pause() function is allowed to be executed by an authorized person as it is restricted with an onlyAuthorized modifier. Both the admin and owner addresses are authorized to call this function. Also, the precondition to call this function is that the state must not have been previously paused, as it is protected with the whenNotPaused modifier.

Let's see when the access restriction pattern should be used.

Applicability

The access restriction pattern should be used in the following cases, when:

- Some functions should only be allowed to be executed from certain roles
- Similar kinds of roles and access are needed for one or more functions or actions
- You want to improve the security of the contracts from unauthorized function calls

Emergency stop pattern

The emergency stop design pattern allows the contract to pause and stop the function calls that could harm the contract state or funds present in the contract. As we know that Ethereum smart contracts are immutable once deployed on the blockchain, it might be possible that the contract can have bugs after deployment and could be found by an attacker to gain the control over the contract or funds. To handle these emergency situations, this design pattern could be helpful to reduce the damage to the contract.

It is important to note that when an emergency stop is activated on a contract, all its stakeholders must be able to see the current state of the contract. This should be done to ensure that the stakeholders get the correct status of the contract as they trust the owner of the contracts (maybe a company in this case).

The emergency stop or pause should only be in control of the owner or authorized person. Only they are allowed to call these functions. However, pausing or stopping the contract could potentially add trust issues to the contracts. You should avoid using this pattern; however, in cases when centralized control is needed, these functions can be added to improve the security of the funds and contracts. For example, the project called **Wrapped BTC (WBTC)** has WBTC tokens on Ethereum blockchain. Each WBTC token is pegged to one **bitcoin (BTC)**. This project has used pausing functionality to pause the token transfer. The owner of the contract is allowed to pause the contract in extreme circumstances. You can see its code present on GitHub at https://github.com/WrappedBTC/bitcoin-token-smart-contracts/blob/master/contracts/token/WBTC.sol.

In the following code sample, we have a contract called Deposit. This contract holds the funds deposited to this contract:

```solidity
contract Deposit {

    address public owner;
    bool paused = false;

    modifier onlyOwner() {
        require(msg.sender == owner);
        _;
    }

    modifier whenPaused() {
        require(paused);
        _;
    }

    modifier whenNotPaused() {
        require(! paused);
        _;
    }

    constructor() public {
        owner = msg.sender;
    }

    function pause() public onlyOwner whenNotPaused {
        paused = true;
    }

    function unpause() public onlyOwner whenPaused {
        paused = false;
    }
```

```
function deposit() public payable whenNotPaused {
    //Ether deposit logic here
}

function emergencyWithdraw() public onlyOwner whenPaused {
    owner.transfer(address(this).balance);
}
}
```

As you can see, the owner of the contract is allowed to pause or unpause the contract:

- The owner can call the pause() function to pause the contract, which, in turn, stops the deposit of the funds and allows onlyOwner to call the emergencyWithdraw() function.
- The owner can call the unpause() function to unpause the contract. This puts the contract back to normal and should allow the deposit of funds again.

Let's see when and where the emergency stop pattern should be used.

Applicability

The emergency stop design pattern should be used in the following cases, when:

- You want your contract to be handled differently in case of any emergency situations
- You want the ability to pause the contract functions in unwanted situations
- In case of any failure, you want to stop contract failure or state corruption

Creational patterns

Creational patterns are used to create different contracts as and when needed. There are some situations when a new contract creation is required from within a contract. The factory contract pattern is the most-used design pattern for creating new contracts. The creation pattern also helps a contract to only create a predefined contract. Let's look at the factory contract pattern.

Factory contract pattern

The factory contract pattern is used when you want to create a new child contract from a parent contract. The benefit of this pattern is that the contract definition is predefined in the main contract and the new contract is deployed with this predefined code. This protects the new contract code from an attacker, as the code of the contract is already defined.

In the following example code, we have a LoanMaster contract. This contract has a createLoanRequest() function that can be called by anyone. This further calls the createInstance() function of the LoanFactory contract to deploy a new instance of a Loan contract using new Loan(). Here, the new solidity keyword is used to create and deploy a new instance of the Loan contract and call its constructor with the given arguments. The creator of the Loan contract is assigned as a borrower in the Loan contract, as shown in the following sample code:

```
contract LoanMaster {
    address[] public loans;
    address public loanFactoryAddress;

    function createLoanRequest(address _token, uint _loanAmount) public {
        address loan = LoanFactory(loanFactoryAddress)
            .createInstance(msg.sender, _token, _loanAmount);
        loans.push(loan);
    }
}

contract LoanFactory {
    function createInstance(
        address _borrower,
        address _token,
        uint _loanAmount
    ) public returns (address) {
        return new Loan(_borrower, _token, _loanAmount);
    }
}

contract Loan {
    address token;
    address borrower;
    uint loanAmount;

    constructor(address _borrower, address _token, uint _loanAmount) public
    {
        borrower = _borrower;
        token = _token;
        loanAmount = _loanAmount;
```

```
    }

    //Other logic of Loan contract
}
```

The `LoanFactory` contract is deployed separately from the `LoanMaster` contract. This way, the `LoanFactory` contract can be used by any of the contracts to create any number of new `Loan` contracts. In the preceding example, a `Loan` contract is used between a single borrower and a single lender.

Let's understand when and where factory contract patterns should be used.

Applicability

The factory contract pattern should be used in the following cases, when:

- A new contract is required for each request to be processed. For example, in the case of creating a new loan term between two parties, a master contract can create a new child contract called `Loan`. This new `Loan` contract has logic to handle contract terms and conditions along with the funds as well.
- You would need to keep the funds separate in a different contract.
- A separate contract logic should be deployed per request and one or more entities are to be tied together using this newly deployed contract.

Behavioral patterns

The behavioral design patterns are used when a contract behaves differently according to its current state. In Solidity, we can have a contract in different states based on the values of its variables. These behaviors should be defined in the contract definition. According to the behavior defined in a contract, a contract transitions between different states. Let's look at some of the behavioral design patterns.

State machine pattern

In a state machine, the initial state and the final state are known for the machine/process. Between these two states, there could be more intermediate states. In each state, different behaviors or functions are allowed on the machine. For example, a vending machine is a state machine, in which it first asks for the user to select the item number, and, once selected, it moves to the collect amount state. Once the amount is paid by the user, it dispenses the item requested by the user and, at last, the process ends.

The state machine pattern allows a contract to transition from different states and enables certain functions to be executed in each state.

A Solidity contract can be designed in such a way that it represents the state in which it transitioned. Based on the state of the contract, you can allow or disallow functions present in the contract. Using the state machine pattern, you can define different states according to your needs and transition the contract from its initial state to its final state. During this transition, the processing of the contract can allow or disallow its functions, so it can successfully transition to its next state.

In the following code, we have a `LoanContract` example. As you can see, it has an `enum LoanState`, which represents different states in which a contract can transition. Let's understand the meaning of the different states:

- `NONE`: No state represented
- `INITIATED`: When a loan is initiated by `borrower`
- `COLLATERAL_RCVD`: When collateral is received from `borrower` in the contract
- `FUNDED`: When the loan received the required collateral from `borrower` and `lender` funded the loan
- `REPAYMENT`: When `borrower` accepted the funding and starts repaying the loan
- `FINISHED`: When the loan has finished, and `borrower` has paid off the full loan

The following is the code for the `LoanContract` contract:

```
contract LoanContract {
    enum LoanState { NONE, INITIATED, COLLATERAL_RCVD, FUNDED,
        REPAYMENT, FINISHED }

    address public borrower;
    address public lender;
    IERC20 token;
    uint collateralAmount;
    uint loanAmount;
    LoanState public currentState;
```

```
    modifier onlyBorrower() {
        require(msg.sender == borrower);
        _;
    }

    modifier atState(LoanState loanState) {
        require(currentState == loanState);
        _;
    }

    modifier transitionToState(LoanState nextState) {
        _;
        currentState = nextState;
    }

    constructor(IERC20 _token, uint _collateralAmount, uint _loanAmount)
        public transitionToState(LoanState.INITIATED) {
        borrower = msg.sender;
        token = _token;
        collateralAmount = _collateralAmount;
        loanAmount = _loanAmount;
    }

    function putCollateral()
        public
        onlyBorrower
        atState(LoanState.INITIATED)
        transitionToState(LoanState.COLLATERAL_RCVD)
    {
        require(IERC20(token)
            .transferFrom(borrower, address(this), collateralAmount));
    }
    //Rest of the code
}
```

In the preceding code, we have an atState() modifier, which is used to check the current state of the loan. The transitionToState modifier is used to transition the contract state from one state to another. As you can see in the putCollateral() function, we used the atState(LoanState.INITIATED) modifier. This allows the function call only in the INITIATED state. Once the function call is executed successfully, it transitions from the INITIATED state to the COLLATERAL_RCVD (collateral received) state.

Let's understand when and where the state machine pattern can be applied.

Applicability

The state machine pattern should be used in the following cases, when:

- A contract needs to transition from different states
- A contract needs to allow different functions and to behave differently during each of the intermediate states

Iterable map pattern

The iterable map pattern allows you to iterate over the mapping entries.

In Solidity, you can define a mapping. The mapping holds the key-value pairs. However, there is no way to iterate over the mapping entries in Solidity. There are some cases for which you would need to iterate over the mapping entries; for these situations, you can use this pattern.

Note that the iteration over the mapping entries should not cause an out-of-gas exception. To avoid these situations, use the iteration only in the `view` function. This way, you would execute a function via message calls only, without generating a transaction.

In the following code, we have `DepositContract`, in which anyone can deposit ether. When ether is deposited, its entry is updated in the `balances` mapping to track how much ether is deposited. It also adds an entry into the `holders` address array, which is used to maintain the unique list of addresses that have been deposited in the contract:

```
contract DepositContract is Ownable {

    mapping(address => uint) public balances;
    address[] public holders;

    function deposit() public payable {
        require(msg.value > 0);
        bool exists = balances[msg.sender] != 0;
        if (!exists) {
            holders.push(msg.sender);
        }
        balances[msg.sender] += msg.value;
    }

    function getHoldersCount() public view returns (uint) {
        return holders.length;
    }
}
```

As you can see in the preceding code; in the `deposit()` function a new depositor address is added in `holders` array. Using this `holders` array you can iterate over the `balances` mapping

Using the `holders` address array, you can find the count of depositors in the contract by calling the `getHoldersCount()` function. Similarly, if you need to filter out some data mapping, you can do it by writing a `view` function.

Applicability

The iterable map pattern should be used in the following cases, when:

- You need iterable behaviors over the Solidity mappings
- There would be fewer mapping entries that would require iterable behavior
- You would need to filter some data out of the mapping and get the result via a `view` function

Indexed map pattern

The indexed map pattern allows you to read an entry from a map using an index. The pattern also allows you to remove an item from a map and an array.

As we have seen in the previous section, in an iterable map pattern, an item is added to a map as well as to an array. Using the iterable map, we can fetch records using an index. However, the iterable map pattern is useful when you just want to keep on adding items in a map and in an array; it does not support the removal of an item. Hence, when you want to remove an item from the map and array, things get a little tricky. Here, in the `IndexedMapping` library contract, we can maintain the list of items and remove them from the list as well:

```
library IndexedMapping {
    struct Data {
        mapping(address=>bool) valueExists;
        mapping(address=>uint) valueIndex;
        address[] valueList;
    }

    function add(Data storage self, address val) internal returns (bool) {
        if (exists(self, val)) return false;

        self.valueExists[val] = true;
        self.valueIndex[val] = self.valueList.push(val) - 1;
```

```
        return true;
    }

    function remove(Data storage self, address val) internal returns
    (bool) {
        uint index;
        address lastVal;

        if (!exists(self, val)) return false;

        index = self.valueIndex[val];
        lastVal = self.valueList[self.valueList.length - 1];

        // replace value with last value
        self.valueList[index] = lastVal;
        self.valueIndex[lastVal] = index;
        self.valueList.length--;

        // remove value
        delete self.valueExists[val];
        delete self.valueIndex[val];

        return true;
    }

    function exists(Data storage self, address val) internal view
    returns (bool) {
        return self.valueExists[val];
    }

    function getValue(Data storage self, uint index) internal view
    returns (address) {
        return self.valueList[index];
    }

    function getValueList(Data storage self) internal view returns
    (address[]) {
        return self.valueList;
    }
}
```

We have defined the `IndexedMapping` contract as a library so that it can be used in any contract. As you can see, we have a `Data` struct to maintain the data for this library. In the `Data` struct, we have the following variables:

- `valueExists`: This is a map from the `address` type to the `bool` type. This map entry tells us whether a particular address exists in the list or not. When an address exists, its `bool` value will be set to `true`, otherwise, it will be set to `false`.
- `valueIndex`: This is a map from the `address` type to the `uint` type. This map entry stores the address to array index mapping.
- `valueList`: This is an array of addresses. This stores the addresses in the list.

As you can see, there are two functions (`add()` and `remove()`) to add and remove an item from the address list. The `getValue()` function is used to ensure an item is present at the specified index in an array.

Let's see where the indexed map pattern can be applied.

Applicability

The indexed map pattern can be used in the following cases, when:

- You want indexed access of an element in a single operation (order of 1 $O(1)$ operation), instead of iterating an array of elements. This is only when the iterable map feature is also required.
- You also want indexed access for elements along with support to remove elements from the list.

Address list pattern

The address list pattern is used to maintain a curated list of addresses by the owner.

In contracts, there are some situations in which you would need a curated list of addresses. For example, you would need a list of whitelisted addresses that are allowed to call a certain function of your contract. Another example is when you want to maintain a list of supported tokens addresses to allow your contracts to interact with these selected tokens only.

In the address list pattern, adding and removing addresses from the list can only be done by the owner of the contract. In the following code, we have an `AddressList` contract:

```
contract AddressList is Ownable {

    mapping(address => bool) internal map;

    function add(address _address) public onlyOwner {
        map[_address] = true;
    }

    function remove(address _address) public onlyOwner {
        map[_address] = false;
    }

    function isExists(address _address) public view returns (bool) {
        return map[_address];
    }
}
```

As you can see in the code, only the owner can add or remove addresses in the contract by calling the `add()` or `remove()` functions. The `isExists()` function is a `view` function and is open for anyone to call. Another contract can even call this function to check whether an address is present in this list or not.

Let's understand where the address list pattern should be applied.

Applicability

The address list pattern should be used in the following cases, when:

- You want to maintain a curated list of addresses.
- You want to maintain the whitelisted address, which is allowed/disallowed to perform a certain task.
- You want to maintain a list of contract addresses that are allowed. For example, an address list of ERC20 token contract addresses.

Subscription pattern

The subscription pattern is used when you need to provide a periodic subscription fee for any kind of service.

A contract can provide different kinds of features or a premium service. Any subscriber can subscribe for services that your contract provides. To enable this, you can charge a subscription fee for a period of time from the subscriber. To implement this service, you can create a Subscription contract, as we have developed in the following code example:

```
contract Subscription is Ownable {
    using SafeMath for uint;
    //subscriber address => expiry
    mapping(address => uint) public subscribed;
    address[] public subscriptions;
    uint subscriptionFeePerDay = 1 ether;

    modifier whenExpired() {
        require(isSubscriptionExpired(msg.sender));
        _;
    }

    modifier whenNotExpired() {
        require( ! isSubscriptionExpired(msg.sender));
        _;
    }

    constructor(uint _subscriptionFeePerDay) public {
        subscriptionFeePerDay = _subscriptionFeePerDay;
    }

    function isSubscriptionExpired(address _addr) public view returns
    (bool) {
        uint expireTime = subscribed[_addr];
        return expireTime == 0 || now > expireTime;
    }

    function subscribe(uint _days) public payable whenExpired {
        require(_days > 0);
        require(msg.value == subscriptionFeePerDay.mul(_days));
        subscribed[msg.sender] = now.add((_days.mul(1 days)));
        subscriptions.push(msg.sender);
    }

    function withdraw() public onlyOwner {
        owner().transfer(address(this).balance);
    }

    function useService() public whenNotExpired {
        //User allowed to use service
    }
}
```

As you can see, in this `Subscription` contract, at the time of deployment, we have configured subscription-per-day fees. This fee is charged per day to the subscriber. The subscriber will call the `subscribe()` function to subscribe for the service for a given number of days. To do this, they have to send the ether along with the function call. After a successful subscription, they will be able to call the `useService()` function to use the service until it expires. Once the service of a subscriber expires, they can renew their subscription by calling the `subscribe()` function again. The owner of the contract can withdraw subscription payments from the contract at any time.

Let's discuss when this subscription pattern can be used.

Applicability

The subscription pattern can be used in the following cases, when:

- You have a periodic paid service-based model
- A user has an existing request and they want to purchase a premium status for their request for a limited period of time
- An existing contract can purchase a premium status for a limited duration

Gas economic patterns

Ether has an economic value and is being traded on exchanges. Ether is used as a crypto fuel to execute transactions on the Ethereum blockchain. In Solidity, each function execution consumes gas. The gas consumed is always paid in ether from the transaction initiator to the block miner. Higher gas consumption by a contract would need more ether; similarly, lower gas consumption would incur a lower amount of ether, and thus lowers the execution cost. Hence, it is always preferred to write the contract in such a way that it can consume the least amount of gas possible for the processing of each transaction.

One possible way to reduce gas consumption is to always deploy the contract with an enabled optimization flag. This process optimizes the EVM bytecode, which then consumes less gas.

We will discuss some of the patterns that can be used to reduce the gas consumption of the contract while carrying out certain types of operations.

String equality comparison pattern

In Solidity, there is no native function to compare the strings. Hence, it is not possible to compare the equality of the two strings. There are ways to do this by checking byte-by-byte data, but it would be a costly operation in terms of gas consumption. Solidity is not ideal for this, so you should think carefully when using this.

To check for string equality, we can generate the keccak256 hash of both strings and compare the hashes with each other. The generated hash of the two same strings will always give the same hash value. If the strings are not equal, their hashes will also differ.

The following code should be used to compare two strings for their equality:

```
function compare(string memory a, string memory b) internal returns (bool)
{
    if(bytes(a).length != bytes(b).length) {
        return false;
    } else {
        return keccak256(a) == keccak256(b);
    }
}
```

As you can see, in the compare() function, first, it checks that the length of the strings are equal, then generates the hash, and compares them. If the length of the strings are not equal, it returns false. If the length of both of the strings are equal, then generate the keccak256 hash of both strings and check. The result of the comparison is returned.

Let's look at when and where a string equality pattern should be used to reduce gas.

Applicability

The string equality comparison design pattern should be used in the following cases, when:

- You want to compare two different strings for equality.
- The string length is larger than the two characters.
- There could be multiple sizes of strings passed to a function and we want to have the gas-optimized solution.

Tight variable packing pattern

Tight variable packing should be utilized when using structs in Solidity. The Solidity language allows structs in the contract, that is used to define an abstract data type. When the storage is allotted to a struct type variable in EVM storage, it is allotted in slots. Each storage slot is 32 bytes long. When statically sized data types are used in the struct (for example, `uintX`, `intX`, and `bytesX`), these variables are allotted storage slots starting from 0 index. Storing and reading data from these storage slots consumes gas based on the number of storage slots written or accessed. Hence, when variables are not tightly packed in a struct, it could consume more storage slots, which would result in more gas consumption during each function call.

Consider the following example code in which we have a `Record` struct consisting of four fields:

```
contract StructPacking {

    struct Record {
        uint param1; // 1st storage slot
        bool valid1; // 2nd storage slot

        uint param2; // 3rd storage slot
        bool valid2; // 4th storage slot
    }

    Record[] records;

    //Params can be passed to function
    function addRecord() public {
        Record memory record = Record(
            1, true,
            2, true);

        records.push(record);
    }
}
```

When the preceding contract is deployed and the `addRecord()` function is executed, it consumes 122,414 gas for the transaction cost and 101,142 gas for the execution cost. The EVM allotted four storage slots for the `record` variable to store. Let's understand how the storage slots are allotted for each of the variables present in the `Record` struct:

- `param1`: This variable is of the `uint` type (that is, `uint256`), using 256 bits, meaning 32 bytes. Hence, it would consume a full storage slot.

- valid1: This variable is of the bool, type, which consumes 1 byte. As the first slot is fully consumed in storing param1, EVM would allocate the second storage slot and store the valid1 variable in it.
- param2: This variable is of the uint type, which would also require 32 bytes of storage. As the valid1 variable does not consume a full storage slot, 31 bytes are still free; however, 32 bytes (256 bit) of data cannot be accommodated in 31 bytes storage. Hence, EVM would allocate a new third storage slot for the param2 variable and store it.
- valid2: This variable is of the bool type, which consumes 1 byte. However, the third storage slot is fully occupied by param2, hence valid2 would be stored in a new fourth storage slot.

In total, four storage slots are allotted to store a Record struct variable.

You can optimize the struct by reorganizing it and consuming fewer storage slots. Solidity does not perform automatic reorganization to tightly pack struct variables. Hence, this has to be done manually.

In the following code, we have reorganized the variables present in the Record struct:

```
contract StructPacking {

    struct Record {
        uint param1; // 1st storage slot
        uint param2; // 2nd storage slot

        bool valid1; // 3rd storage slot
        bool valid2; // 3rd storage slot
    }

    Record[] records;

    //Params can be passed to function
    function addRecord() public {
        Record memory record = Record(
            1, 2, true, true);

        records.push(record);
    }
}
```

The Solidity optimizer does not optimize `structs`.

When the preceding contract is deployed, an `addRecord()` function is called; it consumes 102,219 gas as the transaction cost and 80,947 gas for the execution cost. If you compare the gas cost difference from a previous execution, this new code with a tightly packed struct consumes less gas by 20,195 gas-per-function call. This is a considerably high gas saving per function call. If this function is called multiple times with the preceding tightly packed struct, you would be able to save a lot of gas units.

Let's understand how storage slots are allotted by EVM using optimized code:

- `param1`: This variable is of the `uint` type (that is, `uint256`), using 256, bits meaning 32 bytes. Hence, it would consume a full slot.
- `param2`: This variable is of the `uint` type, which would also require 32 bytes of storage. There is no empty space left in the previous storage slot. Hence, EVM would allocate a new second storage slot for the `param2` variable and store it.
- `valid1`: This variable is of the `bool` type, which consumes 1 byte. As the second storage slot is fully consumed in storing `param2`, EVM would allocate the third slot and store the `valid1` variable in it.
- `valid2`: This variable is of the `bool` type, which consumes 1 byte. However, the third storage slot is not fully occupied, hence, the `valid2` variable will be stored in the third storage slot.

This way, the new optimized code would consume only three storage slots in storing a variable of the `Record` type struct.

Let's understand when the tight variable packing pattern should be used for a struct.

 A `bool` variable in Solidity takes 1 byte to store its value.

Applicability

A tight variable packing pattern is only applicable to `struct` types. It is the developer's responsibility to check the structs used in each contract and ensure that they are tightly packed. Tight packing should be used only when you are certain that a lot of gas could be saved by optimizing the struct. Otherwise, if it is not making any significant difference, you can avoid it.

Life cycle patterns

A Solidity contract is created either by deploying a new contract or by creating from within a contract. Apart from these, a contract and its functions can follow a different life cycle. A Solidity contract can also be destroyed by calling the `selfdestruct` function. Once a contract is destroyed, it cannot be recreated on the same address.

Similarly, using the timestamp's global variables and time units, contract states can also be moved. It is also possible to allow the contract functions to be called based on these timestamps. Let's look into the life cycle design patterns.

Mortal pattern

The mortal pattern allows a contract to be destroyed from the Ethereum blockchain. As you know, a Solidity contract can be destroyed using a `selfdestruct()` function call in the contract. You can have this function call in your contract to allow it to be destroyed once the contract job is over. Once a contract is destroyed, the contract states would not remain on the blockchain, and if there are any ethers present in the contract, it would be sent to an address passed to the `selfdestruct()` function as argument.

As you can see in the following sample code, there is a `kill()` function. Only the owner of the contract can call this function:

```
import "openzeppelin-solidity/contracts/ownership/Ownable.sol";

contract Mortal is Ownable {

    //...
    //Contract code here
    //...

    function kill() external onlyOwner {
        selfdestruct(owner());
```

```
        }
    }
```

Once the owner of the contract calls the `kill()` function, the contract is destroyed from the Ethereum blockchain and any ether present in the contract is sent to the owner.

> The mortal pattern is very dangerous when used in the contract. You should not use this pattern in the first place.

Let's understand when and where the mortal pattern should be used.

Applicability

The mortal pattern should be used in the following cases, when:

- You do not want a contract to be present on the blockchain once its job is finished
- You want the ether held on the contract to be sent to the owner and the contract is not required further
- You do not need the contract state data after the contract reaches a specific state

Auto deprecate pattern

The auto deprecate pattern allows time-based access to certain function calls. In Solidity, using the `block.timestamp` and `now` calls, you can get the time of the block when it is mined. Using this time, you can allow or restrict certain function calls in Solidity. There are also some globally available time units present in the Solidity language that you can use. These time units are `seconds`, `minutes`, `hours`, `days`, `weeks`, and `years`. You can use these time units to calculate the Unix epoch time in the past or in the future.

In the following code, the `contribution()` function should only be called in a specified duration of time:

```
contract AutoDeprecate {

    uint startTime;
    uint endTime;

    modifier whenOpen() {
        require(now > startTime && now <= endTime);
        _;
```

```
    }
    constructor(uint _startTime, uint _endTime) public {
        require(_startTime > now);
        require(_endTime > _startTime);
        require(_endTime > _startTime + 1 weeks);
        startTime = _startTime;
        endTime = _endTime;
    }
    function contribute() public payable whenOpen {
        //Contribution code here
    }
    function isContributionOpen() public view returns(bool) {
        return now > startTime && now <= endTime;
    }
}
```

The contribution() function uses the whenOpen modifier. The whenOpen modifier only allows the function call between the startTime and endTime variables. Both of these times are set from the constructor of the contract and the duration between these two times must be equal to or greater than a week.

Let's look at when and where auto deprecate patterns should be used.

Applicability

The auto deprecate design pattern should be used in the following cases, when:

- You want to allow or restrict a function call before or after a specified time
- You want to allow or restrict a function call for a specified duration of time
- Auto-expire a contract, which would not allow any function calls after the expiry time

Summary

In this chapter, we learned about different Solidity design patterns that can be used while developing contracts. The design patterns should be used according to your contract's architecture needs. We learned about security, creational, behavioral, economic, and life cycle patterns. Under each category of patterns, we discussed different patterns. We looked at the access restriction, state machine, subscription, tight variable packing, and auto deprecate pattern.

Solidity is a contract-oriented language, hence there are some different kinds of design patterns applied and used. The blockchain ecosystem is relatively new and more such design patterns could evolve over time. It is also possible that the Solidity language itself natively evolves and supports some of the design patterns.

When developing a contract code, it is better to use these design patterns so that it is easy to maintain the contract and ensure it is readable for other developers, so they can understand the logic quickly. However, a developer should not restrict themselves from using only these design patterns; they can also use different patterns in conjunction with these patterns as and when needed.

In the next chapter, we will talk about the tips, tricks, and security best practices for Solidity contract writing.

Questions

1. Do withdrawal patterns also apply to ERC20 tokens?
2. Should we always use the access restriction pattern?
3. When should the emergency stop pattern be used?
4. Should new contract creation always be done using the factory contract pattern?
5. Does enabling the optimization flag tightly pack structs?
6. Should the mortal pattern always be used?
7. Can the time used for the auto deprecate pattern be manipulated?

14
Tips, Tricks, and Security Best Practices

In `Chapter 2`, *Getting Started with Solidity*, and `Chapter 3`, *Control Structures and Contracts*, we learned about the Solidity language and its control structures. However, there are a few things that a developer should know while writing smart contracts.

In this chapter, we will look at the best practices of writing smart contracts in Solidity. These best practices will help developers to write bug-free code. Also, there have been many attack patterns that can cause damage to funds present in the smart contracts in production. We will discuss these attacks and how to prevent your contracts from these attacks.

This chapter will cover the following:

- Some of the best practices we should use while writing contracts
- Ethereum on-chain data is public; you should not share any confidential data
- Avoiding untrusted external function calls
- Preventing a front-running attack
- Preventing a reentrancy attack
- Preventing a signature replay attack
- Preventing integer overflow and underflow attacks

Technical requirements

The code related to this chapter can be found at this GitHub location: `https://github.com/PacktPublishing/Mastering-Blockchain-Programming-with-Solidity/tree/master/Chapter14`.

You will need the following versions of the tools installed:

- Truffle v5.0.10 or later
- Solidity v0.5.0 or later
- Node v8.11.3 or later

Smart contracts best practices

In Solidity, there are many global variables and constructs available for developers to use. However, there are some limitations you must know about, otherwise, it could harm your contracts in production.

If your contract has bugs or incorrect architecture, which allows an attacker to gain unauthorized control over your contracts, an attacker can steal funds from your contracts, or they can perform unintended operations on contracts, which, in turn, can harm you or your users economically. There have been many attacks in past that have allowed an attacker to gain full control over a contract and steal millions of dollars worth of ether. One such example is the Parity MultiSig wallet hack, in which more than 150,000 ETH was stolen in July 2017.

The Ethereum blockchain is slow in terms of transaction execution. This slowness also creates some problems related to the transaction reordering. Also, the Ethereum blockchain is public, meaning that any transaction data that you passed in is visible to everyone. Hence, you will have to keep those things in mind when you design your contracts.

While writing the smart contracts in Solidity as a developer, you will have to be aware of a few things and constructs. Let's discuss the best practices that you will have to be aware of.

Avoiding floating pragma

The first line of a Solidity contract always starts with defining the `pragma solidity` version. This indicates the Solidity compiler version, using which the code should be compiled. The `pragma` is defined as follows:

```
// Bad Practice
pragma solidity ^0.5.0;
```

The preceding code signifies that the code should be compiled with any Solidity version starting from `0.5.0` to `0.5.x` because the version starts with a `^` (caret sign). For example, if the latest compiler version of Solidity available is `0.5.8`, then the contract can be compiled with any compiler version between `0.5.0` and `0.5.8`.

It is always recommended that `pragma` should be fixed to the version that you are intending to deploy your contracts with. The contracts should be deployed with the version that they have been tested with. Using floating `pragma` does not ensure that the contracts will be deployed with the same version.

It is possible that if you used floating `pragma` while deploying your contracts, then the most recent compiler version will be picked. The most recent version of compiler has higher chances of having bugs in it. Hence, it is better to keep your contract to a fixed compiler version.

Fixed `pragma` is defined as follows:

```
// Good Practice
pragma solidity 0.5.0;
```

Using the preceding line, you are instructing the compiler to compile the contract only with the `0.5.0` Solidity compiler. The fixed `pragma` also prevent contracts from not using the very latest version of the compiler. As the latest version of the compiler might have some unknown bugs.

Avoid sharing a secret on-chain

Ethereum is a public blockchain; hence, all the transaction data is visible to everyone. To execute a function on a contract, a user has to sign the function data from their EOA and send it to the Ethereum blockchain. In this process, the signed data contains the first four bytes of the function signature followed by the function arguments. This data is visible, and anyone can see the function definition as well, if the code of the contract is published on the block explorer.

If you have a contract that requires confidential data to be sent, it would be visible to everyone, and would not remain private when a transaction is initiated. As the Ethereum blockchain processes transactions slowly, you can see the transaction data and can initiate another transaction. For example, in the game Rock-Paper-Scissors, two players each select one of three options at random, and one wins the game. But if player 2 knows the option chosen by player 1, then player 2 can select the right option so that he/she wins the game all the time:

```
contract RockPaperScissor
{
    enum Choice {NONE, ROCK, PAPER, SCISSOR}

    struct PlayerChoice {
        address player;
        Choice choice;
    }

    PlayerChoice[2] players;

    function registerPlayer() public {
        if(players[0].player == address(0))
            players[0].player = msg.sender;

        if(players[1].player == address(0))
            players[1].player = msg.sender;

        revert("All players registered");
    }

    function play(Choice _choice) public {
        uint index = validateAndfindPlayerIndex();
        players[index].choice = _choice;
    }

    function checkWinner() public {
        //Code to check winner and reward
    }

    function validateAndfindPlayerIndex() internal returns (uint) {
        if(
            players[0].player == msg.sender &&
            players[0].choice == Choice.NONE
        ) return 0;

        if(
            players[1].player == msg.sender &&
            players[1].choice == Choice.NONE
```

```
    ) return 1;

    revert("Invalid call");
    }
  }
```

As you can see in the preceding sample code of the `RockPaperScissor` contract, when the `play()` function is called by player 1, their choice is published on the chain, and player 2 can see player 1's choice. Now player 2 can select a choice so that they always win the game.

To overcome this problem, you can use the commit-and-reveal scheme, in which all the players first share a secret hash, and, once all the players have submitted the secret hash, they all reveal their secret by sharing the salt in the reveal transaction.

The commit-and-reveal scheme

In the commit-and-reveal scheme, first, a hash of the original secret is submitted to the blockchain. This secret hash is recorded and stored on-chain in the contract. Once all the players or parties have submitted their secret hash, they all have to reveal their choice by submitting salt, using which they have generated the secret hash. A **salt** is like a password; using this an user can generate a secret hash. This secret is generated by combining the data (to be hidden) and salt and taking hash of this combined data. You can use any hashing algorithm to generate hashes. We have used `keccak256` hashing algorithm in the sample code

This way of first committing the secret hash and later revealing it prevents players from sharing their original choice. Let's look at the updated code of `RockPaperScissor`:

```
contract RockPaperScissor
{
    //Rest of the exiting code

    struct PlayerChoice {
        address player;
        bytes32 commitHash;
        Choice choice;
    }

    function play(bytes32 _commitHash) public {
        uint index = validateAndFindPlayerIndex();
        players[index].commitHash = _commitHash;
    }
```

```
function reveal(Choice _choice, bytes32 _salt) public {
    require(
        players[0].commitHash != 0x0 &&
        players[1].commitHash != 0x0
    );
    uint index = findPlayerIndex();
    require(players[index].commitHash ==
        getSaltedHash(_choice, _salt));
    players[index].choice = _choice;
}

function getSaltedHash(Choice _answer, bytes32 _salt)
    internal view returns (bytes32) {
    return keccak256(abi.encodePacked(address(this), _answer, _salt));
}
}
```

In the preceding code, using the play() function, a player is submitting his secret commit hash. Once all the players have submitted their commit hashes, each player should call the reveal() function to reveal their choices.

There are some guidelines you must follow for salt usage:

- The salt that you have revealed on-chain must not be used again in future transactions. It must be different each time.
- The salt must be strong enough in terms of number of characters used, so that it becomes difficult to brute-force.
- If you have used salt while testing on the testnet chain, you should not use the same salt again on the mainnet chain.
- You must keep you salt stored at secret location until it's revealed.

Be careful while using loops

You can use loops in two ways as bounded or unbounded loops. If you are performing some operations in contract or just calculating some results, you can use either of the loops.

You can have loops for any function in the Solidity language. However, if the loop is updating some state variables of a contract, it should be bounded; otherwise, your contract could get stuck if the loop iteration is hitting the block's gas limit. If a loop is consuming more gas than the block's gas limit, that transaction will not be added to the blockchain; in turn, the transaction would revert. Hence, there is a transaction failure. Always consider having bounded loops in which you are updating contract state variables for each iteration. Try to avoid using loops that change contract state variables in the first place.

You can have unbounded loops for view and pure functions, as these functions are not added into the block; they just read the state from the blockchain when message calls are made to these functions.

However, if these view or pure functions (containing loops) you are using in other public/external functions, it could block your contract operation because the view or pure functions would consume gas when they are being called from non-pure / non-view functions. Let's look at the following code:

```
contract DividendCalculator is Ownable {
    struct Account {
        address payable investor;
        uint dividend;
    }
    Account[] investors;
    mapping(address => Account) investorsMap;

    modifier onlyInvestor() {
        require(msg.sender == investorsMap[msg.sender].investor);
        _;
    }

    function calculateDividend() public onlyOwner {
        //Bad Practice
        for(uint i = 0; i < investors.length; i++) {
            uint dividendAmt = calcDividend(investors[i].investor);
            investors[i].dividend = dividendAmt;
        }
    }

    function withdraw() public onlyInvestor {
        Account memory account = investorsMap[msg.sender];
        uint dividendAmount = account.dividend;
        account.dividend = 0;
        account.investor.transfer(dividendAmount);
    }

    function calcDividend(address investor) internal returns
    (uint){
        //Logic to calculate dividend
    }
}
```

As shown in the preceding code, the owner can calculate the dividend amount for each investor. However, in the `calculateDividend()` function, there is an unbounded loop, which could start failing the transaction because of insufficient gas, once it starts consuming more gas units than the block gas limit.

To avoid these issues, you must pass in the number of iterations a loop can execute:

```
//Good Practice
function calculateDividend(uint from, uint to) public onlyOwner {
    require(from < investors.length);
    require(to <= investors.length);
    for(uint i = from; i < to; i++) {
        uint dividendAmt = calcDividend(investors[i].investor);
        investors[i].dividend = dividendAmt;
    }
}
```

The preceding code would help the owner of the contract create multiple batches to calculate dividend. This avoids the transaction failure issues related to insufficient gas.

Avoid using tx.origin for authorization

The `msg.sender` global variable gives the address of the caller of the function. The `tx.origin` is also a globally available variable that returns the address of the transaction initiator. For example, using an EOA account; Alice initiates a transaction to Contract-A which further makes a function call to a Contract-B. Then the function present in Contract-B would give the address of the Contract-A when `msg.sender` is evaluated; however, when `tx.origin` is evaluated in the same function, it would return the address of Alice's EOA, because Alice is the original transaction initiator.

The `tx.origin` method should not be used as authorization for any function. The access control protected using `tx.origin` can be attacked and would allow an attacker to gain unauthorized access rights.

In the following code, we have a `Vault` contract, which keeps the ether of an owner. The owner of the `Vault` contract can withdraw their ether at any point in time:

```
contract Vault {
    address authorized;

    modifier onlyAuthorized() {
        // Bad Practice
        require(authorized == tx.origin);
        _;
```

```
    }

    function withdraw(address beneficiary) public onlyAuthorized {
        beneficiary.transfer(address(this).balance);
    }
}
```

An attacker writes the `AttackerContract` code as follows:

```
contract AttackerContract {
  address targetContract;
  address attackerWallet;

  function () external payable {
    Vault(targetContract).withdraw(attackerWallet);
  }
}
```

Here is the diagram explaining how an attacker could attack his victim:

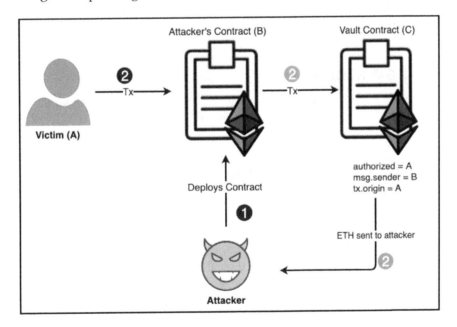

Attacker steals ETH from Victim's contract via tx.origin

In the preceding diagram, we have shown how an attacker attacked on the Vault contract which is using tx.origin in the onlyAuthorized modifier, in order to check the authorization of the function caller. The following actions take place in this scenario:

1. An attacker will create an AttackerContract contract and deploy it. Somehow, an attacker would ask the original owner of the Vault contract to send some ether to the AttackerContract contract.

2. Once the original owner sends the ether to AttackerContract, transaction calls the Vault.withdraw() function. Then, it would check that tx.origin is the authorized person of this contract, execute the withdraw() in the Vault contract, and send all the ether present in the Vault contract to the attacker's wallet.

We have discussed an attack in which there was a **Victim** (user) and two contracts involved. However, this is not the only case where this kind of attack is possible. There could be many more contracts in between as well. Also, you need to know that when you are using an external library contract, all the function calls are made using delegate calls to the library functions.

Preventing an attack

You should not use tx.origin in your contract to check for the authorization. Always use msg.sender to check the authorization of the function calls:

```
modifier onlyAuthorized() {
    //Good Practice
    require(authorized == msg.sender);
    _;
}
```

In the preceding code, we are using msg.sender instead of tx.origin. This fixes the issue, and an attacker would not be able to perform the attack because msg.sender always returns the address of the previous caller in the call stack. In the previous attack scenario, msg.sender would return the address of AttackerContract; hence, authorization fails, and funds would not be sent to the attacker.

The timestamp can be manipulated by miners

In the Solidity language, there are two globally available variables: `block.timestamp` and `now`. Both are aliases and are used to get the timestamp when the block is added to the blockchain. The timestamp is in Unix epoch time in seconds. This timestamp is set by the miner when solving the **Proof-of-Work (PoW)** puzzle. The timestamp of the new block must be greater than the timestamp of the previous block. However, the miner can manipulate the timestamp.

If your contracts need random numbers, you should not use `block.timestamp` or `now`, as these can be manipulated by miners. Consider the following example:

```
// Bad Practice
function random() public view returns (uint){
    bytes32 hash = keccak256(
        abi.encode(blockhash(block.number), block.timestamp));
    return uint(hash);
}
```

As you can see, the random number generation is using `blockhash` and `block.timestamp`. As `block.timestamp` can be manipulated by miners, they can set it in a certain way so that they can benefit from it. The random number generation from the contract is still not very robust; hence, you should not generate random numbers in contracts.

The 15-second blocktime rule

According to the yellow paper of Ethereum, the timestamp for the new block must be greater than the previous block; otherwise, the block will be rejected. The timestamp can be manipulated by the miners using the client software they are running to mine the blocks. There are two Ethereum client software used mostly, and these are **Geth** and **Parity** clients. Using their software code and algorithms, these two clients maintain the 15-second rule such that the timestamp difference between the two blocks should not be more than 15 seconds. Otherwise, the client software rejects the block.

Hence, if you are using `block.timestamp` or `now` in your contract code, you need to ensure that you are not performing any processes that would require less than 15 seconds.

Carefully making external function calls

The Solidity smart contracts do have a size limit for deployment. A contract that consumes less than 8 million gas at the time of deployment can be deployed on the Ethereum blockchain at the moment. In the future, this limit of 8 million gas could be increased. When your contracts are big enough and cannot be deployed in a single block, they need to be further divided into multiple small contracts. You can break a big contract into multiple small contracts and libraries and interconnect these contracts via function calls with the required authorization.

Not only that, if your contracts needed to communicate with some external service, you would need to make external function calls to external contracts. For example, the **Oraclize** service is a third-party service that provides APIs to fetch data from the internet and lets you use it in the blockchain. The Oraclize service has exposed some external function calls, which your contracts can call and use.

When you need to call external functions, the developer must know under which category the target contract belongs to, such as the following:

- **Trusted contracts**: A contract that is deployed and managed by you is known as a **trusted contract**. External function calls to trusted contracts often not create any issues, as they are known.
- **Untrusted contracts**: A contract that is deployed and managed by another entity, for example, Oraclize contracts. These are known as **untrusted contracts**. External function calls to these untrusted contracts might have some security implications in future if their contracts are attacked.

However, the definition of an untrusted contract is up to the developers and the project; if they believe that certain external service contracts can be treated as trusted contracts, they can use it. If they are not sure about the contract code or the third-party contract, then they can classify that contract as an untrusted contract.

To give an example, if your contracts require integration with external contracts, such as KyberNetwork's decentralized exchange, you can use them as KyberNetwork contracts are security audited and have been used by many people for a considerable amount of time. You can take KyberNetwork contracts as trusted. If you are unsure about some external contract and the code is not open, then you can treat these contracts as untrusted contracts. However, once again this is your decision when architecting the design of your contracts and integration with external services.

Let's discuss some of the things that should be avoided when making external function calls from a contract.

Avoid dependency on untrusted external calls

As we have discussed, there are some contracts or services that are managed by an external third party. The risk in calling an external function on these services is very high. Our external function calls are dependent on their contract and code, and hence it might be possible for these external services to inject malicious code; if executed, your contracts would behave unexpectedly. It is always recommended to have fewer untrusted external function calls.

Ensure that enough due diligence is done while choosing an external contract for integration with your contracts.

Avoid using delegatecall to untrusted contract code

By using the `delegatecall` function present in the contract, you can refer some code from another contract and execute it on the current contract context. Library functions are delegate-called to the current contract execution.

When the target contract address is untrusted, and/or when you are not sure about the code, you should not make `delegatecall` to these untrusted contracts. These untrusted contracts can perform malicious operations on your contracts as follows:

```
function _delegate(address _target) internal {
    // Bad Practice
    _target.delegatecall(bytes4(keccak256("externalCall()")));
}
```

As you can see from the preceding code, although the _delegate function is `internal`, it still takes the `_target` argument. If the `_target` contract address is an untrusted contract, it can perform any arbitrary code execution on your contract. If the target contract is killed via `selfdestruct`, the external call to the function will always fail, and if there is any dependency of your contract on that target contract, your contract would stuck forever.

To avoid such untrusted calls, only use trusted contract addresses to perform the `delegatecall` operation. Also, do not allow users to pass in the _target contract address.

Rounding errors with division

In the Solidity language, there is no official support for floating-point numbers. In the future, there will be support for floating-point numbers. Developers have to use unsigned `uintX` or signed `intX` integers only for their calculations. Hence, if any division operation is performed, the result of that calculation might have rounding errors. The result is always rounded down to the nearest integer.

Here is an example in the following code:

```
//Bad Practice
uint result = 5 / 2;
```

The preceding code would set value 2 in the `result` variable, as 2.5 rounded down to the nearest integer.

These rounding errors could cause problems when you are calculating some values that affect tokens, bonuses, or dividends. Hence, a developer must know that the rounding errors would be possible in the calculations and must ensure the best way possible to mitigate these issues:

```
uint bonusTokens = balances[msg.sender] * dividendAmount / totalSupply;
```

As you can see from the previous code, there could be some rounding errors present in the `bonusTokens` calculation.

To prevent rounding errors to a certain extent, you should do the multiplication (if multiplication exists) of the values first and then only perform the division operation, as this would have less rounding errors. Otherwise, performing calculations on variables that have some rounding errors already could cause more issues in the final result:

```
//Bad Practice
uint intermediateResult = 5 / 2;
uint finalResult = 10 * intermediateResult;

//Good Practice
uint finalResult = (10 * 5) / 2;
```

As you can see, the first `finalResult` variable will be assigned with a value of 20. However, the second `finalResult` variable will be assigned with a value of 25 as we performed multiplication before the division operation.

Using assert(), require(), and revert() properly

In Solidity, three functions are provided by the language to check for the invariant, validations, and to fail the transaction. These functions are as follows:

- `assert()`: The `assert()` function should be used when you want to check for invariants in the code. When any invariant is incorrect, the code execution stops, transaction fails, and contract state changes are reverted. This function should only be used for invariant checking. It should not be used for input validation or pre-condition checking.
- `require()`: The `require()` function should be used when you want to validate the arguments provided to the function. It is also used to check for the valid conditions and variable values to be in an expected state. If the validation fails, the transaction also fails, and the contract state changes are reverted.
- `revert()`: The `revert()` function should be used to simply fail the transaction. Ensure that the `revert()` function is called under some certain conditions. Once this function is called, the transaction fails, and the contract state changes are reverted. This should be used when you cannot use the `require()` function.

Gas consumption

In the Ethereum blockchain, every contract deployment and transaction consumes gas. The gas is paid in ether and has an economic value. When you deploy a contract, the compiled code is sent to the blockchain. The bigger the contract code is, the more gas units it consumes when deployed. It is recommended that the contract should be compiled with optimization flag enabled. Enabling optimization might optimize the code size and consume less gas units when deployed. Sometimes, the optimized contract also consumes less gas units when their function is called.

You can also look at the *Economic patterns* section in `Chapter 13`, *Solidity Design Patterns*, to learn some special cases where gas consumption could be optimized.

Known attack patterns

On the Ethereum blockchain, many hacks occurred between the years 2017 and 2018, and they continue happening because of buggy contract code. In the year 2018 only, more than $1 billion worth of ether and tokens got stolen from Ethereum smart contracts due to vulnerabilities present in the code that attackers exploited. These hacks happened because of bad coding practices and a lack of testing of the contracts.

Looking at how those hacks happened, there have been many attack patterns that have been identified. It is the developer's responsibility to check their code for all of these known attack patterns. If these attack patterns are not prevented, there could be a loss of money (ether/tokens) or the attacker could enforce unintended transactions on the contracts.

Always keep yourself updated with the new features added in the Ethereum hard-forks or soft-forks. These new features could help you reduce the attack surface. However, some new features in Ethereum could allow more new attack patterns. As and when you learn about a new attack pattern, check your contracts and ensure that your contracts do not form these patterns.

Let's look at some of the well-known attack patterns.

Front-running attacks

The Ethereum blockchain is slow, and it is a public blockchain. Because it is a public blockchain, all the transaction data is open and can be seen by others. Even when a transaction is in the pending state, its data can be seen by others.

A transaction can be processed slowly or quickly, depending on the quantity of the gas fees each transaction is going to give to the miner to execute and add to the blockchain. A transaction paying higher gas fees are picked up by the miners first and added to the blockchain. On the other hand, if a transaction is paying lower gas fees, they are picked up later when miners are free and not many higher gas fees transactions are in the pending state. In other words, your transaction is chosen to be executed based on the higher gas fees a miner can earn for executing it. When a transaction is not processed due to low gas fees, it will remain in the pool and be in the pending transaction state.

Once a transaction is in the pending state, all the blockchain clients sync up their pending transactions as well; hence, a pending transaction is also known to each and every client node. You can write a program or script that would listen for any transactions initiated from a specific wallet or with specific parameters. One can also find all pending transactions and see its transaction data.

In a front-running attack, an attacker sees transaction A (which is in pending state), and they immediately send transaction B with a higher gas price to get economic benefit.

Let's discuss the front-running attack that is possible with a specific implementation of ERC20's `approve()` function.

Example of an ERC20 approve function

In the ERC20 token standard, there is a function called `approve()`, following is the implementation code for it. You can also refer `Chapter 7`, *ERC20 Token Standard*, *The approve function*, for more details on the working of the code:

```
function approve(address spender, uint tokens)
public returns (bool success)
{
    allowed[msg.sender][spender] = tokens;
    Approval(msg.sender, spender, tokens);
    return true;
}
```

This function is always prone to a front-running attack if not handled correctly. Let's see how a front-running attack happens on the `approve()` function, with the help of the following diagram, showing a transaction flow step by step:

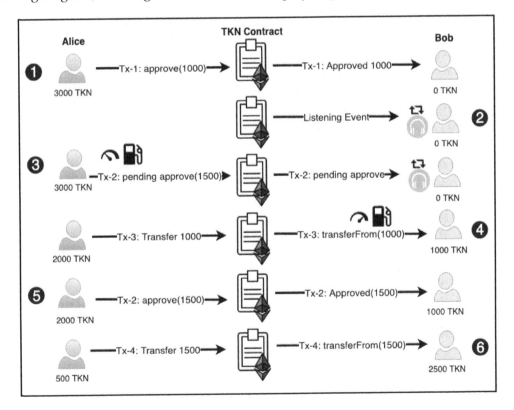

Front-running attack transaction flow

In the preceding diagram we have the following:

- There are two people named **Alice** (left) and **Bob** (right)
- There is an ERC20 token contract (symbol: **TKN**) on which both Alice and Bob interact and initiate transactions
- Initially, **Alice** holds **3,000 TKN** and **Bob** holds **0 TKN**
- We are assuming that both have sufficient ether present in their wallets, in order to initiate transactions
- The sequence of actions are numbered from 1 to 6 in the preceding diagram

Let's go through each numbered action and transactions happened between both **Alice** and **Bob**:

1. **Alice** initiates a transaction (*Tx-1*) and calls the `approve(Bob, 1000)` function on the TKN contract to give approval of 1,000 TKN tokens to **Bob**. This transaction is executed, confirmed, and **Bob** is approved for 1,000 TKN.

2. **Bob** starts listening for events on the blockchain. He keeps listening for an event when any transaction from **Alice** is initiated. He also gets **Alice** in confidence that she approved fewer tokens; originally, he wanted to have 1,500 TKN approved. It could also be that **Alice** and **Bob** communicated and agreed—before *Tx-1* happened—that **Bob** actually needs 1,500 TKN; however, by mistake, **Alice** approved only 1,000 TKN.

3. Now, **Alice** realized the mistake, initiated a transaction (*Tx-2*), and calls the `approve(Bob, 1500)` function to approve 1,500 TKN tokens to **Bob**.

4. **Bob** immediately got the notification from his event listener that **Alice** initiated the new *Tx-2* transaction (the transaction is still in the pending state) to give him approval for 1,500 TKN. However, before *Tx-2* gets confirmed, **Bob** initiated transaction *Tx-3* (to front-run *Tx-2*), and calls the `transferFrom(Alice, Bob, 1000)` function to transfer 1,000 TKN from **Alice**'s account to **Bob**'s account as he already has 1,000 TKN approved. **Bob** initiate this transaction with the higher gas price to get his *Tx-3* transaction confirmed and executed before transaction *Tx-2*. For example, transaction *Tx-2* is initiated with a gas price of 21 gwei. **Bob** will initiate his *Tx-3* transaction with a gas price of more than 21 gwei, for example, 40 gwei. We assume that **Bob's** transaction *Tx-3* gets confirmed and executed before Alice's transaction, *Tx-2*. Hence, transaction *Tx-3* would transfer 1,000 TKN from Alice's account to Bob's. After this transaction, **Alice** has 2,000 TKN and **Bob** has 1,000 TKN in their respective wallets. Also, Bob's approved balance would be 0 TKN after transaction *Tx-3*, as he has used the approval and transferred 1,000 TKN tokens.

5. Now, Alice's *Tx-2* transaction got confirmed and executed. This transaction gave **Bob** a fresh approval of 1,500 TKN.

6. As **Bob** is already listening for events, he gets a notification that transaction *Tx-2* is also confirmed and executed. He immediately initiates another transaction, *Tx-4*, and calls the `transferFrom(Alice, Bob, 1500)` function. Once this *Tx-4* transaction is confirmed, **Alice** would have 500 TKN left in her wallet, and **Bob** would have managed to get 2,500 TKN by performing a front-running attack.

As we have seen in the preceding example; originally **Alice** wanted to approve **1,500 TKN** tokens only, however **Bob** end up getting **2,500 TKN** tokens.

We discussed how a front-running attack can be used by an attacker to transfer more tokens than intended. Let's discuss the front-running prevention techniques.

Preventing an attack on the approve function

To overcome an attack on the `approve` function, there are different techniques.

For the previously discussed problem of the `approve` function, one solution is to set the allowance to 0 (zero) before setting it again with a new value:

```
function approve(address spender, uint tokens)
    public returns (bool success)
{
    require((tokens == 0) || (allowed[msg.sender][spender] == 0));
    allowed[msg.sender][spender] = tokens;
    Approval(msg.sender, spender, tokens);
    return true;
}
```

The preceding code would prevent the front-running attack and would enforce that the approver would always set the allowance to 0 (zero) before setting it again with a new non-zero value. However, this technique requires two transactions in case the approver wants to change the allowance.

Here is another technique in which two extra functions are provided by the contract. These functions would allow the approver to increase or decrease the allowance when required. The code for these functions is as follows:

```
function increaseAllowance(address spender, uint256 addedValue) public
returns (bool) {
    require(spender != address(0));
```

```
        _allowed[msg.sender][spender] =
            _allowed[msg.sender][spender].add(addedValue);
        emit Approval(msg.sender, spender, _allowed[msg.sender][spender]);
        return true;
    }

    function decreaseAllowance(address spender, uint256 subtractedValue) public
    returns (bool) {
        require(spender != address(0));

        _allowed[msg.sender][spender] =
            _allowed[msg.sender][spender].sub(subtractedValue);
        emit Approval(msg.sender, spender, _allowed[msg.sender][spender]);
        return true;
    }
```

The increaseAllowance() function would allow the approver to increase the allowance by the provided number of tokens. However, the decreaseAllowance() function would allow the approver to decrease the allowance. Using these functions, the front-running attack is prevented. You can also refer Chapter 7, *ERC20 Token Standard, Advance functions*, for more details on these functions.

Other front-running attacks

We only discussed the approve() function-specific front-running attack. However, other kinds of front-running attacks can also happen. For example, when a user is registering a unique value, once this is registered, no one is allowed to register it again on the same contract. Like the domain name registration, once it is registered with a user, another person cannot register it again, as the first person has became the owner of that.

An attacker can watch for the transactions on that contract and can send the high gas-price transaction to front run the user's transaction.

To prevent this type of front-running attack, you should use the commit-and-reveal scheme. We discussed this technique in this chapter when we were discussing about not sharing confidential information on chain in the *Avoid sharing a secret on-chain* section.

This type of attack is mostly dependent upon how you write your contract code. If your contract is vulnerable to front-running attacks and have some ether fund movements linked to it, an attacker could attack your contracts more often to gain more ether.

Anyone is allowed to initiate a transaction with a high gas price specified in the transaction. For this, they can watch the current network's high gas price and send a transaction with a higher gas price than that. For example, the current networks' high gas price is X wei. They can send a transaction with $X + Y$ wei to get their transaction added to the block as soon as possible.

Reentrancy attacks

Many hacks in the past have used this technique. In this attack technique, an attacker deploys a new contract and calls a specific function on the target contract. The call sends ether to the attacker's contract, and their contract makes a function call to the target contract again. This process continues in a loop until all the ether or funds from the target contract is drained in the attacker's contract.

Let's look at an example where a reentrancy attack is possible:

```
//Bad Practice
function withdraw() public {
    uint amount = balances[msg.sender];
    msg.sender.transfer(amount);
    balances[msg.sender] = 0;
}
```

In the preceding code, using the `withdraw()` function, a user can withdraw their ether balance, which they have deposited to this contract previously. As you can see in the code, it reads the user's balance and sends that amount of ether to the function caller, and, at the end, it resets the balance of that function caller. As we learned previously, in Solidity, you can write a fallback function that can receive ether and execute some code. An attacker can deploy a contract, in which they will add a fallback function, as follows:

```
//Code used by an Attacker
address attackedAddress = 0x1234;

function attack() public onlyOwner {
    attackedAddress.withdraw();
}

function() external payable {
    while(attackedAddress.balance > 0) {
        attackedAddress.withdraw();
    }
}
```

In the preceding code `attackedAddress` value is the address of the contract (which contains ether) on which an attack will be performed. An attacker will deploy this contract and initiate attack to a target contract via the `attack()` function:

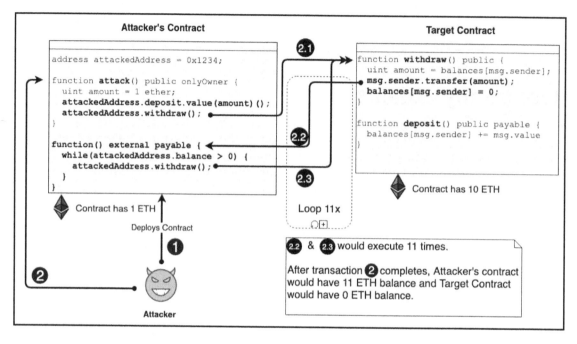

Reentrancy attack by an attacker

The preceding diagram shows how an attacker could exploit reentrancy vulnerability present in the **Target Contract**. Let's look at the transaction flow shown in the preceding diagram:

Lets assume the following is the initial state of the contract:

- **Target Contract** has 10 ETH present in it.
- **Attacker's Contract** has not deposited any ETH to **Target Contract** yet. It means the attacker's contract, `balance[msg.sender]`, is 0 in **Target Contract**.

Transactions are executed in the following order:

1. The attacker deploys his contract called an **Attacker's Contract**. He also deposits 1 ETH into his deployed contract. This transaction is shown as **(1)** in the preceding diagram.

2. The attacker calls the `attack()` function (transaction **(2)** shown in the diagram) on his deployed contract and the following internal transactions are executed:

    1. The `deposit()` function call deposits 1 ether to **Target Contract**. This updates the **Attacker's Contract** `balance[msg.sender]` balance to 1 ether. It further calls the `withdraw()` function in the **Target Contract**. This internal transaction is shown as **(2.1)** in the preceding diagram.

    2. The `withdraw()` function present in **Target Contract** sends 1 ether to **Attacker's Contract** via the `msg.sender.transfer(amount)` function call. This, in turn, triggers the fallback function of **Attacker's Contract**. This transaction is shown as **(2.2)** in the preceding diagram.

    3. The fallback function checks that **Target Contract** still has some ether balance in it. If it has balance left, then make a call to the `withdraw()` function in the **Target Contract**. This transaction is shown as **(2.3)** in the preceding diagram.

3. The preceding process of internal transactions **(2.2)** and **(2.3)** continues until **Target Contract's** ether balance is not empty.

4. After executing **(2.2)** and **(2.3)** 11 times, transaction **(2)** would complete and the balance of **Attacker's Contract** would be 11 ether, and the balance of **Target Contract** would be 0 ether.

As you can see using the above reentrancy attack an attacker was able to drain the target contract. The conditions and loop settings might vary according to the code of the attacker's contract and the target contract.

Let's look at the technique to prevent a reentrancy attack.

Preventing a reentrancy attack

To prevent a reentrancy attack, the state of the variables should be updated first, and then ether should be sent to a user's account as follows:

```
// Good Practice
function withdraw() public {
    uint amount = balances[msg.sender];
    balances[msg.sender] = 0;
    msg.sender.transfer(amount);
}
```

In the preceding code, we are updating the balance of the user's account to 0 (zero), and then only sending the ether to the user.

Remember to always update the relevant state variables first and then only transfer ether at the last step.

Replay attack

This is a type of attack in which an attacker is allowed to recall the function of the contract, allowing them to update the state variables. Using these attacks, an attacker can update the state variables or perform some unintended operations multiple times when they should not be allowed. The signature replay attacks are mostly prone to replay attacks. You should ensure that signatures are handled correctly in contracts. Let's discuss the signature replay attack.

Signature replay attacks

There are some cases when a user signs some data off-chain and the data is given to some other authorized user who will submit the signed data on the contract. This process allows users to perform transactions even when off-chain and later, the confirmation or trade is updated on-chain. For example, in projects such as 0xProject, where trades are matched off-chain by signing the order data and later on, actual trade is updated on-chain.

Let's look at an example:

```
import "openzeppelin-solidity/contracts/cryptography/ECDSA.sol";

contract ReplayAttack {
    using ECDSA for bytes32;

    //Bad Practice
    function submitRequest(
        address _signer,
        address _target,
        uint _param1,
        uint _param2,
        bytes memory _signature
    )
        public onlyAuthorized
    {

        bytes memory input = abi.encode(_target, _param1, _param2);
        bytes32 inputHash = keccak256(abi.encodePacked(input));
```

```
        inputHash = inputHash.toEthSignedMessageHash();
        address recoveredAddress = inputHash.recover(_signature);

        require(recoveredAddress == _signer);

        //Further action on target address
        _target.submitRequest(_param1, _param2);
    }
}
```

In the preceding code, a user signs the data related to the `submitRequest()` function and sends it to an authorized user off-chain; later, the authorized person submits the signature along with the signed data to the `submitRequest()` function. The function checks the inputs signed by the user themselves; otherwise, the transaction will fail.

However, the preceding code is prone to signature replay attack because the same signed data can be sent again by the authorized person. Sending this data again, they can perform unintended operations that are expected to be performed only once.

Preventing a signature replay attack

To prevent a replay attack, you should use the **nonce** in the signed data. The user should sign the data along with a unique nonce value each time. Also, the nonce should be stored on-chain, to show that the user has previously sent the signature with that nonce:

```
mapping (address => mapping(uint => bool)) nonceUsedMap;

function submitRequest(
    address _signer,
    address _target,
    uint _param1,
    uint _param2,
    uint _nonce,
    bytes memory _signature
)
public onlyAuthorized {

    bytes memory input = abi.encode(_target, _param1, _param2, _nonce);
    bytes32 inputHash = keccak256(abi.encodePacked(input));
    inputHash = inputHash.toEthSignedMessageHash();
    address recoveredAddress = inputHash.recover(_signature);

    require(recoveredAddress == _signer);
    require(nonceUsedMap[_signer][_nonce] == false);

    nonceUsedMap[_signer][_nonce] = true;
```

```
          //Further action on target address
          _target.submitRequest(_param1, _param2);
      }
  }
```

As you can see in the preceding code, we have introduced a mapping that takes the address of the signer and the nonce used by the signer to sign the data. The call is executed when the nonce is not used previously. The transaction fails when a nonce is used previously.

There are other prevention methods you can also use to prevent a signature replay attack, as follows:

- If there are multiple functions in your contract that accept the exact same type of signed data, in that case, a user signs the data thinking that function A() will be called by an authorized person. However, an authorized person can also call function B(), as both of the functions needed the same type of parameters to be signed. To prevent these types of attacks, you can include the function name in the signed data. Having this, even an authorized person cannot call incorrect functions. To include a function name in the signature data, you could also use msg.sig (this gives us the first four bytes of the function called).

- You can also maintain the incremented sequence of the nonce on-chain in the contract itself. This ensures that the function execution from an authorized person can only be performed in sequence according to the nonce increment. This is a good solution when you need only the sequential execution of the transactions. This way, you do not need to track the nonce off-chain on the client side. Also, the nonce would not go off-sync as it starts from 0 and keeps increasing by 1 with every signed transaction. This also lets the user know what is the last nonce they have used, all the other nonce numbers after that are still not used or valid. For example, if a user has sent 50 transactions and the last nonce used is 49, they know that from 50 onwards, all the nonce values are still valid and have not been used.

In the upcoming Ethereum hard fork named **Istanbul**, there would be a new instruction to get the chain ID of the network in the contract. The chain ID is the fixed and unique ID associated with each testnet and mainnet out there. Adding this chain ID to your signature data would ensure that no one can reply to your signatures, which were previously used on testnet and played on mainnet.

Integer overflow and underflow attacks

In Solidity, there are data types that represent fixed-size integers. For example, all uint8 datatype from uint8, uint16, and moving up to uint256, increasing the bit value by 8 each time. Each type has a different limit to store an integer. For example, a variable of type uint8 can store values from 0 to 28-1 (0 to 255).

Similarly, int8 up to int256 (increasing the bit value by 8 each time) are also prone to integer overflow or underflow attacks.

When a value of the variable reaches the upper limit and further increases, it will cause integer overflow and the value goes back to zero. Also, when the value of the variable reaches the lower limit and further decreases, it will cause integer underflow, and the value goes back to a maximum value of the data type.

For example, you have an int8 variable in the contract. The value assigned to it is 255 (the maximum value an int8 variable can hold). Now, you increase the value of this variable just by 1, either using arithmetic operators such as +, +=, or ++. The new value of the variable would be 0 (the lowest value an int8 variable can hold) as it has caused integer overflow.

The values of these variables are increased or decreased using some operators. These operators are as follows:

- Arithmetic operators: +, −, and *
- Arithmetic and assignment operators: +=, −=, and *=
- Pre and post, increment, and decrement operators: ++ and −−

You need to be cautious when using the preceding operators for arithmetic operations:

```
// Bad Practice
function transfer(address _to, uint256 _value) public {
    balanceOf[msg.sender] -= _value;
    balanceOf[_to] += _value;
}
```

The preceding transfer() function is prone to both integer overflow and integer underflow attacks. Because there is no check present that ensures that the _value variable can have a valid value that would not cause integer overflow or integer underflow. An attacker can pass a high value for the _value argument so that the balance of msg.sender is increased to the maximum of uint256; that way, they can perform an integer underflow attack. Similarly, they can decrease the balance of _to to zero.

One way to avoid integer overflow or underflow attacks in Solidity code is to check for the boundaries of the data type before assigning new values; however, this can be dangerous if any condition is missed. Doing this requires extra care and attention while writing code:

```
// Good Practice, but not the best as more code is required to prevent
// from integer overflow and underflow attacks
function transfer(address _to, uint256 _value) public {
    require(balanceOf[msg.sender] >= _value);
    balanceOf[msg.sender] -= _value;
    balanceOf[_to] += _value;
}
```

Instead, you can use the `SafeMath` library provided by the OpenZeppelin libraries. This library reverts the transaction when integer overflow or underflow happens:

```
import "openzeppelin-solidity/contracts/math/SafeMath.sol";

contract ERC20 {
    using SafeMath for uint;
    mapping(address => uint) balanceOf;

    // Good Practice
    function transfer(address _to, uint256 _value) public {
        balanceOf[msg.sender] = balanceOf[msg.sender].sub(_value);
        balanceOf[_to] = balanceOf[_to].add(_value);
    }
}
```

In the preceding code, we are using the `SafeMath` library; when using the `sub()` or `add()` library functions, we do not even need to check for other conditions as the transaction would revert automatically when the integer overflow or underflow happens.

Note that the `SafeMath` library is for the `uint256` data type only. For the `int256` data type, you can use the `SignedSafeMath` library present in the OpenZeppelin libraries. You can refer Chapter 9, *Deep Dive Into the OpenZeppelin Library, Math-related libraries*, for more details on these library files.

Ether can be sent forcibly to a contract

While writing a contract, you can define a `payable` fallback function to accept ether in your contract, as follows:

```
function() external payable {
}
```

If this fallback function is not present in a contract and it does not have the `payable` modifier for any function, then your contract is not meant to receive ether.

However, there is still a possible way to send ether to a contract that does not accept ether. This is possible via a `selfdestruct` function call:

- Let's assume there is a contract *x* that has some ether present in it.
- Also, there is a contract *y*.
- Contract *x* calls the `selfdestruct(address_Of_ContractY)` function.
- This process sends all ether present in contract *x* to contract *y*, even if contract *y* neither has a fallback function nor a `payable` function.

The code snippet from contract *x* is as follows:

```
function kill(address _contractY_addr) external onlyOwner {
    selfdestruct(_contractY_addr);
}
```

The preceding is the code present in contract *x*; once the `kill()` function is called by the owner of the contract *x*, all the ether present in this contract will be sent to contract *y*. Even If contract *y* has a `payable` fallback function defined; in this case of forcible sending of ether, it will not be executed, however, the ether balance of contract *y* would increase silently.

By using the previous approach, an attacker could affect the behavior of your contract in certain situations: if your contracts are only accepting ether from authorized sources and via either fallback or `payable` functions, and if your contract code contains some decision logic using `address(this).balance` (this gives the current ether balance of the contract). Then, an attacker can influence the decision logic, as he would be able to manipulate the contract's ether balance using an unauthorized method.

Prevention and precaution

At the moment, there is no possible way to prevent forceful ether sending from happening.

However, the developer should caution that you should not assume any specific amount of ether in the contract and write the contract logic. If you do so, an attacker can attack and might lock your contract by sending some ether to your contract forcefully. For example, you should not use `address(this).balance` in your code to check the contract ether balance and compare it with any specific value.

Security analysis tools

There are some static and dynamic security analysis tools available online that you can also use to find the security-related issues in your contract. It is recommended that you should use these tools during the development and testing processes as well. These tools are as follows:

- **Securify**: An open source, online, and fully automated static analyzer tool for Solidity smart contracts. It scans the contract code for vulnerability patterns and generates a report. You just need to upload your contract code into the online tool and get the report generated in a few minutes. You can use this tool at `https://securify.chainsecurity.com/`.
- **Slither**: An open source Solidity static analyzer that detects many common Solidity issues. The open source code of the tool can be found at `https://github.com/crytic/slither`.
- **Mythril**: An open source security analysis tool for EVM byte codes. It finds security vulnerabilities in smart contracts built on EVM compatible blockchains such as Ethereum, Quorum, Vechain, Rootstock, and Tron. You can find the tool at `https://github.com/ConsenSys/mythril`.
- **MythX**: The platform and the ecosystem for Ethereum security tools. The tool also provides an extension for Truffle and Embark. You can use the online MythX platform at `https://mythx.io/`.
- **SmartCheck**: This open source tool provides static analysis of the Solidity contract, detects vulnerabilities, and checks for best practices. You can find the tool online at `https://tool.smartdec.net/`.

Let's use the Securify tool to find the security vulnerabilities in Solidity contracts.

Using the Securify tool

The Securify tool is a static analyzer tool for Ethereum Solidity contracts. This tool scans the contract code and finds the security vulnerability patterns in the code. After scanning, it generates a report along with descriptions of each vulnerability it has found and provides an idea of how to solve each vulnerability.

It is a free online service that anyone can use; you just need to put in your Solidity code and click on the **SCAN NOW** button to scan your code for vulnerabilities. You can go to `https://securify.chainsecurity.com/` and use the online version of the tool. You can also download its open source code and run the scanner in a Docker container. You can find the code on GitHub at `https://github.com/eth-sri/securify`.

The UI of the tool looks like this:

```
   SCAN NOW        REQUEST AUDIT       DISCORD                    PASTE CODE   UPLOAD ZIP   CLONE GIT

 1  pragma solidity 0.4.25;
 2  library SafeMath {
 3    function add(uint256 a, uint256 b) returns (uint256 c) {
 4      c = a + b;
 5      assert(c >= a);
 6      return c;
 7    }
 8
 9  function mul(uint256 a, uint256 b) internal pure returns (uint256 c) {
10    if (a == 0) {
11      return 0;
12    }
13    c = a * b;
14    assert(c / a == b);
15      return c;
16    }
17  }
18
19  contract Ownable {
20      using SafeMath for uint256;
21
22      address owner;
23
24      modifier onlyOwner() {
25        require(msg.sender == owner);
26        _;
27      }
```

By using Securify, you accept the Terms of Service.

Securify scanner online interface

As you can see, you can paste your code and scan it. You can also upload the ZIP file containing Solidity contract files and get all of them scanned.

Let's use the Securify tool with a sample code. For this, we are using the `Example.sol` code present on GitHub at `https://github.com/PacktPublishing/Mastering-Blockchain-Programming-with-Solidity/blob/master/Chapter14/contracts/Example.sol`. We pasted the code into the Securify online tool and clicked on the **SCAN NOW** button.

The tool generates the following report:

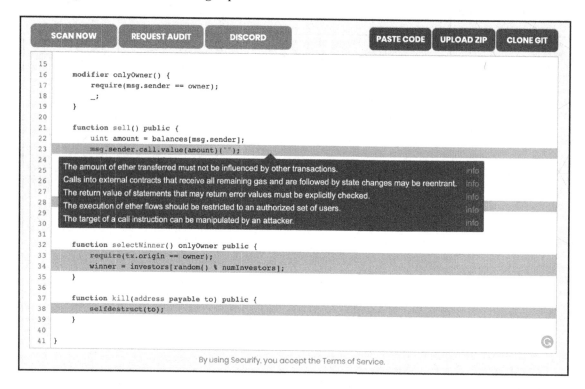

Securify scanned code and generated the report

As you can see, each line shows the problems that have been found. You can click on the **info** line present in each line to get detailed information about each issue.

The tool also generates the following report. If you scroll down in the Securify web page, you would get the detailed report as follows:

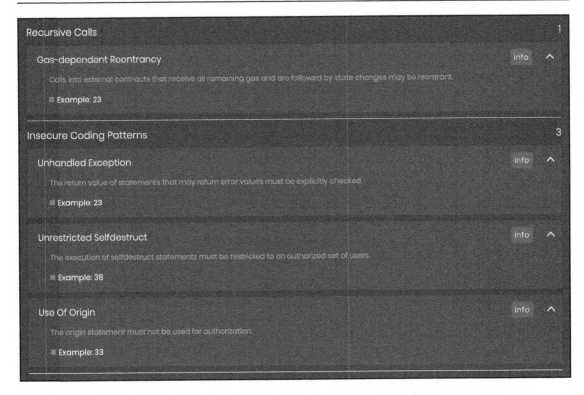

Securify detailed report of vulnerabilities found

You can keep on fixing Securify report issues and scanning your code. At least all the issues marked with red in the preceding screenshot should be fixed to ensure you have non-vulnerable code.

Summary

In this chapter, we covered best practices such as avoiding floating `pragma`, the commit-reveal scheme, using external function calls, and integer rounding errors. Additionally, we discussed attack patterns such as front-running, reentrancy, signature replay attacks, and integer overflow and underflow attacks. These are the most important things to know, as writing contracts in Solidity can be easy, but writing a bulletproof contract is hard.

There have been many hacks, and we have covered some of the most well-known ones, but it's the developer's responsibility to keep checking for newly identified attack patterns, so that they do not make the same mistakes again while writing contracts. Apart from that, always use the latest versions of the libraries. The Solidity compiler does fix the issues in the new versions; keep yourself up to date with the latest changes and ensure that the new bugs found in the compiler are not affecting your contracts.

We started this book by looking at Ethereum and blockchain, as well as discussing Solidity in depth. We also discussed Truffle, Ganache, Remix IDE, MetaMask, and how to use these tools while writing contracts. We learned how to use the ERC20 and ERC721 standards, as well as OpenZeppelin libraries. We looked into contract upgradability and we covered best practices. These topics are the most important to learn for a Solidity contract developer. Keep learning new things, libraries, and tools for the Ethereum blockchain. I would love to see you develop contracts for Ethereum, so keep building and evolving the ecosystem.

Questions

1. If the on-chain data is public, how do you perform private transactions?
2. Should a developer use `delegatecall`?
3. How can you prevent a reentrancy attack?
4. How can you prevent integer overflow and underflow in a contract?
5. How do you use unbounded loops in contracts?
6. How do you perform decimal point calculations?

Assessments

Chapter 1

1. In the first chapter, we discussed many different properties of blockchain. Of those properties, immutability and replacing trust are the most unique. If your application needs these properties, then blockchain should be used; otherwise, it is not recommended that you use blockchain as part of your application.

2. Blockchain is also called distributed database and distributed ledger technology. As it is a distributed database, it offers many benefits, which we discussed in the chapter on the properties of blockchain; however, centralized databases and other kinds of database have their own benefits. For example, centralized databases are fast in transaction processing, they can handle large amounts of data, and keeping each storage cell persistent does not require any kind of fee. Hence, blockchain technology is not going to replace traditional databases.

3. Blockchain technology provides a way to create and code smart contracts. Smart contracts are self-executing laws that are coded in contract terms. You can code contracts that can be defined in a logical sense and need blockchain technology properties. If you cannot do this with your contract, then you should not use blockchain; not every problem can be solved with blockchain.

4. When two entities want to create transactions between each other, but they do not trust each other, this situation would require a trusted middle man or intermediaries that they both trust. These intermediaries take a commission to allow a transaction to happen between the entities. Blockchain technology provides a way to replace intermediaries completely. There are some instances where intermediaries could influence a transaction's outcome, and in some cases, putting too much trust in these intermediaries could harm entities. Hence, replacing these intermediaries with blockchain technology gives more power to the entities, allowing them to conduct transactions between each other without completely trusting each other.

5. Ethereum is a first-of-its-kind programmable blockchain, as blockchain technology is still very new (it's only been around for a couple of years). Ethereum blockchain transactions take a long time to execute, which makes it a slow transaction system. On the other hand, there is a lot of research going on to improve the transaction processing speed of blockchain, as well as other scalability improvements.

6. The gas limit is just like fuel in your car; until you run out of fuel in your car, it keeps on going. In a similar way, for every transaction, you would need to set a gas limit to let the **Ethereum Virtual Machine (EVM)** know how much fuel you have provided for that transaction. If your transaction takes less than the provided gas limit, it will succeed; otherwise, it will fail. The gas price is the price per gas limit unit. This calculates the fee for the transaction, and that fee is paid to the miners to put your transaction on the blockchain.

7. Immutable code provides the guarantee that the code will not be changed and the execution of it will not be influenced by anyone. Not even a hacker or attacker can change the code. An irreversible transaction guarantees that once a transaction is performed, it is there forever and its outcome cannot be changed. Both of these properties of blockchain are the cornerstones of the trust that blockchain provides.

8. Traditional software development and programs work in cycles of Build -> Test -> Deploy, and these cycles keep on going. This way, even if there is a new feature or bug in your application, you can fix and redeploy it again. However, with smart contracts, you only have a single chance to deploy, because once the smart contracts are deployed, they cannot be changed. However, you can perform any number of iterations for Build -> Test.

9. As we saw in the answer to the previous question, it is not possible to change the code of a smart contract once it is deployed. It makes more sense to test your contract thoroughly before deploying it in production.

Chapter 2

1. Both of these functions are available with the `address` data type. The `transfer` call fails even when an internal function call throws an error; however, the `send` function returns the `bool` type as `false` in the event of a failure.

2. If you use the `send` function in your contract, then you would need to explicitly check for the returned `bool` type, which could be missed, which would lead to a higher chance of getting it wrong. Even if there is no more gas left during the `send` call and an out-of-gas exception is thrown, it will return `false` as a result. For these reasons, it is recommended that you use `transfer` rather than the `send` function.

3. It is not recommended that you use the `delegatecall` function. This functions use low-level calls. These calls use the code from another contract, but use the storage of the current contract, thereby making it possible for another contract to change the state of the current contract. Using the `delegatecall` function should be the last resort.

4. As a developer, you should test the gas requirement of your function. You should have the minimum and maximum gas consumption calculated using the test results of the different set of inputs and conditions. Once you know the maximum gas consumption, you can add some extra gas on top of it and use that amount of gas to pass on the target contract using the `gas` function.

5. There is no concrete way to generate a random number in Solidity. Developers might think that `block.timestamp`, `now`, and `blockhash` can be used to provide a source of randomness, but this is wrong, as these values can be manipulated by miners. This means that, to date, there is no fool-proof way to generate a random number in the Solidity language itself.

6. These functions are used to encode the data type and generate the payload that can be sent to the target contract function calls. The `abi.encode` function just uses the data, along with its padding, and generates `bytes`; however, the `abi.encodePacked` function removes the padding—for example, with the `string` and `bytes` data types, it just takes the string data part.

7. The `assert` function is used to check the invariants while you are developing contracts. This helps developers to find errors and fix them during the development phase. The production code should not use `assert` functions. The `revert` function is used to revert the whole transaction. The remaining gas will be refunded back to the transaction initiator when `revert` is executed. The `require` function takes a `bool` parameter, which is mostly used to verify the input variables, allowed values, preconditions, and authorized access.

Chapter 3

1. A Solidity contract can be large enough to be deployed into a single block of Ethereum blockchain. An Ethereum block has a maximum gas limit that, at the time of writing, is 8 million gas units. If your contract takes 8 million or less gas units at the time of deployment, you can deploy your contract on the blockchain. If it takes more gas than 8 million gas units, consider enabling optimization on your contract; this would reduce the amount of gas significantly. If, even after enabling optimization, your contract is still taking more than 8 million gas units, then you would have to break your contracts into multiple contracts and link those contracts together with interfaces.

2. Prior to Solidity version 0.4.22, you could define a function as having the same name as the contract. That function becomes the constructor of the contract; however, this was deprecated in version 0.4.22 and the `constructor` keyword was introduced.

   There have been some incidents in the past where the developers of a contract unintentionally changed the contract name and forgot to change the constructor function name. This change made the constructor function a non-constructor function, meaning that anyone would be able to call that function as many times as they wanted. This introduced a security vulnerability in the contract.

   To reduce the possibility of unintentional errors by the developer of the contract, Solidity changed the way it defined the constructor. Now, from Solidity version 0.4.22 onward, the use of the `constructor` keyword is recommended to define the constructor. From Solidity version 0.5.0 onward, the old approach was removed and defining the constructor using the `constructor` keyword was made an absolute requirement.

3. The definition of the fallback function says that when a function is not found in the contract, then the fallback function is executed if it is present. This means that when a caller initiates a transaction that contains a function signature (or empty transaction data) that does not match with any of the functions present in the contract, the call will be transferred to the fallback function and it will be executed only if the fallback function is present. If the fallback function is not present or does not have public visibility, then the transaction will fail.

   As per this definition, you can use the fallback function to handle function calls that are not found in the contract. Mostly, the fallback function is used for receiving ether in the contract. To receive ether in your contract, the fallback function must have a `payable` function.

4. Solidity is popular because it is a Turing-complete, contract-oriented language and gives developers the ability to use loops in a contract. But it is the developers' responsibility to use these loops carefully, because each iteration of a loop consumes gas.

   If you have a non `view` or `pure` function that is using loops, you need to ensure that it can perform the expected operations in the lowest number of iterations possible to consume less gas. You must find the best and worst case complexity of the loop and calculate the gas consumption for each case.

   When a loop is used in a `view` or `pure` function, there is no limit to the loop iterations, as these executions will be performed by the Ethereum node itself that your wallet is connected to.

5. You can write the `view` function and call these functions without paying gas for the transaction. Using the `view` function, you can write complex search and filter operations on the data and return the results. While writing the contract, you also need to ensure that the contract returns some filtered results if your client-distributed application needs those results for faster decision-making. Once the contract is deployed, you cannot add more filter functions to your contract.

6. Yes, using libraries, you can write your custom function for the data types supported in Solidity. This is because you can write a function in the libraries and the first parameter of the function's argument takes a type. There, you can state which data type you would want to add this function to. After defining the functions in the libraries, you can apply those custom functions to your data type by using the `using <Library> for <DataType>` keyword.

7. You can use `throw` to report an exception while executing logic. It is used to check the argument's valid inputs for a function, as well as in some other cases. If a transaction encountered `throw` while executing code, all of its remaining gas present in the transaction would also be consumed by the EVM and given to the miner of the block. Sometimes, this would consume a lot of gas and remove a significant amount of ether from the wallet of the user who initiated the transaction, as `throw` is used to notify the user that an exception occurred while executing the transaction, but also consumes the remaining gas. In that context, the `throw` keyword is a misnomer. Hence, the `throw` keyword was deprecated in version 0.4.13 and was removed from the language from version 0.5.0 onward. Instead of `throw`, you should use the `revert` keyword, which refunds the remaining gas to the user.

Chapter 4

1. You can create as many accounts as you want in MetaMask. MetaMask supports the **BIP**-44 (short for **Bitcoin Improvement Proposal**) algorithm to create new EOA accounts using the same secret seed phrase. You can keep on creating new accounts in MetaMask by clicking on **Create Account**, and it will generate a new account. Using the same secret seed phrase, you can create the same accounts again in the same order on another MetaMask instance on other browsers or computers.

2. Yes, you can connect MetaMask to your own private blockchain by configuring the **Custom RPC URL** in the MetaMask. To do this, choose **Custom RPC** from the network list; it will ask you the RPC URL to connect to.

3. The MetaMask secret seed is also called a mnemonic. The mnemonic is created from the English word list so that it can be remembered by humans easily. Using these mnemonics, the user can generate their accounts on any computer or on any of the wallets that support Ethereum. You should keep your mnemonic seed stored in a secret place or in a safe, as this represents your wallet's private keys. If you lose these mnemonics, you won't be able to recover your accounts, and will lose the funds kept in the accounts.

4. Yes, the MetaMask browser plugin maintains Ethereum accounts, and each Ethereum account can receive ERC20 as well as ERC721 tokens. You can also configure the address of your ERC20 tokens in MetaMask to track its balance and enable MetaMask to transfer ERC20 tokens via MetaMask.

5. Yes, you can cancel any pending transaction using the latest version of MetaMask, from 6.6.1+ onwards. You can select a pending transaction from the transaction queue list in MetaMask and click on the **Cancel** button to send a new transaction to cancel that pending transaction.

6. First, your contract should be compiled successfully in the Remix IDE. Now you can click on the **Details** button, under the **Compile** tab on the right-hand side panel. It will open up a pop-up window with all the details about the contract. You can navigate to the **ABI** section and copy **ABI** into the clipboard.

7. It is recommended that you use the optimization settings while compiling your contracts. The optimizer reduces the byte code size if possible. This also lowers the gas consumption on each transaction execution.

Chapter 5

1. Ganache GUI provides full UI/UX control, with lots of advanced features that you can configure. The Ganache GUI can be used while you are doing local development; however, the Ganache CLI is the command-line version of Ganache. You can use `ganache-cli` for your local development environment, as well as automated testing or scripting.

2. You can use Ganache as a personal Ethereum blockchain. It is lightweight and easy to set up. However, for a private Ethereum blockchain, you can also use the Geth or Parity blockchain and run it locally.

3. Yes, you can use the Truffle framework for deploying contracts using migration scripts in a production environment (mainnet). Truffle maintains the migration status locally, which means that, in future, you'll be able to add more migration scripts and can get your production contract states updated. Normally, the users should not run Truffle tests on production (mainnet), as this will cost you ether.

4. Yes, the Truffle framework supports Ethereum blockchain. If you have a private Ethereum blockchain network that was built using Geth or Parity clients, you can easily use Truffle for it.

5. Most of the time, it is better to use Infura URLs to connect to the desired network. However, if you need to send a large number of transactions on the network, you might face some issues with Infura. To send large numbers of transactions, it is better to set up your local node (Geth or Parity) so that it is connected to the desired network.

Chapter 6

1. The Solidity linters do report some security issues at the Solidity level. They also help in finding common security issues.

2. No, having 100% contract coverage does not mean that testing is done efficiently. It only suggests that all the possible branches have been covered by the test cases. However, there could be more test scenarios that are not covered by the test cases; they must be included in the test suite by the developers.

3. When you are trying to debug and find out the internal function traces of a specific function, function traces should be used. This will help you understand the transaction flow in the case of complex contract structures.

4. It is good practice to use the Solidity linters for projects. The linters find the common issues that would improve your code quality, and sometimes also report common security issues.

Chapter 7

1. Ether is not ERC20 compliant, and so it does not support functions such as ERC20 which are applied on ERC20 standard token.

2. Let's say that there is a contract that provides a periodic service to subscribed accounts. To subscribe for this service, the contract accepts a certain amount of a specific ERC20 token as a subscription fee. To do this, the contract must have a `public` or `external` function that calls the `token.transferFrom()` function and charges the caller in an ERC20 token. This way, a user can pay for the service using the ERC20 token. Note that you should not transfer the ERC20 token directly to the service contract, as you might lose your tokens.

3. Yes, there are some other token standards that have been proposed, such as ERC223 and ERC777. Some of these are still under consideration, and are not widely used in production. As of today, ERC20 is the most widely used token standard for fungible tokens.

4. Yes, you can write a function in a contract to accept multiple addresses in an array and transfer the tokens to all of these addresses. This way, you can send tokens to multiple addresses in a single transaction. This also consumes less gas compared to sending tokens to each address individually.

5. If some ERC20 tokens are locked into a contract and there is no way to take the ERC20 out of the contract, then those tokens are locked forever. For example, the ERC20 contract of the **OmiseGo** token (symbol: **OMG**) did not restrict the transfer of OMG tokens to the same token contract address, which caused more than 21,000 OMG tokens to be locked in the contract itself. These tokens were sent by mistake from token holders and cannot be recovered from the contract. You can check these locked tokens at https://etherscan.io/address/0xd26114cd6EE289AccF82350c8d8487fedB8A0C07#tokentxns.

Chapter 8

1. The ERC165 standard is used in the ERC721 standard so that it knows which functions are supported by the contract. Similarly, you can use the ERC165 standard in any other standard or contract to let the client know about the functions supported by the standard or the contract.

2. The ERC721 NFT standard should be used when you have some digital assets that are nonfungible. Nonfungibility means that you cannot further subdivide these assets. As an example, digital collectible cards are nonfungible, and each one is different from others.

3. The _mint() and _burn() functions are both internal functions, and so these can be called from the implementing contract. The _mint() function is used to create a new, unique ERC721 NFT token and send it to the given address. On the other hand, the _burn() function is used to burn a given NFT token from its owner's wallet.

4. There is an advanced standard for ERC721, called the ERC1155 multitoken standard. You can find the complete standard definition at https://github.com/ethereum/EIPs/blob/master/EIPS/eip-1155.md. The standard uses the features of both the ERC20 and ERC721 standards in a single ERC1155 standard.

5. In the ERC721 standard, there are three roles that are allowed to transfer a token. These roles are owner, approver, and operator. Any of these roles can transfer an NFT token, given that the role has the required permission to transfer the token.

6. You can identify the difference between the mint and burn operations by looking at the Transfer event parameters. If a token is minted, the from address in the Transfer event would be the address(0) address. However, in the case of a burned token, the to address in the Transfer event would be the address(0) address.

7. The bytes data contains the function call data that is to be further called and executed by the onERC721Received() function in its context.

Chapter 9

1. It is bad practice to copy and paste the OpenZeppelin library code and use it in your project. You must install the OpenZeppelin npm package in your project. You can stick to using a specific version of the library files in your project.

2. You should not make any modifications in the OpenZeppelin library files present in the node_modules folder. These library files are thoroughly tested, and so if you need a modified version of the files, copy it from the library and maintain your version separately in your project.

3. Sometimes, there are requirements for which you do not need some of the functions of the OpenZeppelin library files. In such cases, you can override specific functions in your implementation to always revert a transaction. This is the best way to handle and remove these functions from the library contract files.

4. You can use the mathematical operations (using the +, -, /, *, and % operators) without using the SafeMath library; however, some of the operators can cause integer overflow or underflow when used without operand boundary checks. Hence, it is always recommended that you use the SafeMath library.

5. There is a small difference between the Ownable and Claimable contracts. In the Ownable contract, the owner can transfer the ownership to another address in a single transaction; however, in the Claimable contract, the owner can initiate the transfer of the ownership and the pending owner has to claim for their ownership: only then is the ownership transferred. Hence, in Claimable, two transactions are required to transfer the ownership. If we compare both of the contracts, Claimable is more secure and allows the current owner to reverse the transaction in the event where an incorrect owner address is provided.

Chapter 10

1. There are many multisig wallet contracts available that can be used by the developer and projects in production. Apart from the Gnosis multisig wallet that we discussed in this chapter, there is a new product from Gnosis, called Gnosis Safe (`https://safe.gnosis.io/`). The Gnosis Safe has recently been built and has not been battle-tested; however, Gnosis multisig is used mostly by projects, and had been battle-tested.

2. Yes, using the Gnosis multisig wallet contract you can receive or transfer any type of ERC token standard, such as ERC20 and ERC721 tokens.

3. Yes, a MultiSig-A wallet can also control a MultiSig-B wallet. For this to happen, the MultiSig-B contract must have the MultiSig-A's address as the owner in the owner's list.

4. As of now, if the number of owners in a Gnosis multisig wallet is greater than the number of required signatures, you cannot assign a specific owner to compulsorily sign every transaction. This is the limitation of the Gnosis multisig wallet. In future, there could be other multisig wallets that support this feature.

5. It is always recommended that you have more owners in the multisig than the number of signatures required for any transaction to execute. If this is ensured in a multisig wallet, then you can add a new owner to the multisig and remove the one who lost their private key; however, in cases when the number of owners and the number of required signatures are equal, then you cannot recover your multisig wallet. Hence, the funds that are kept in and the privileges assigned to the multisig wallet are at risk.

6. In the Gnosis multisig wallet GUI, there is a configuration or settings page in which you can configure the email address that will be used to send notifications directly to your email address in the event of any transaction-related events.

7. To sign any transaction, normally, you do it via MetaMask. MetaMask in hot wallet and transaction signing happens on a machine that is connected to the internet. You can set up an offline or air-gapped machine, on which you can sign the transaction, take the signed data, and send it via a machine connected to the internet. Offline signing is more secure, compared to hot wallets.

Chapter 11

1. No, it is only when you require your upgradable contracts that you should use the ZeppelinOS framework; otherwise, it is better to use Truffle.

2. No, you should not have constructors defined for your upgradable contracts, as contract initialization is performed via the `Proxy` contract.

3. To achieve true decentralization, you need to ensure that future rule changes are not allowed in contract. Using the immutable property of blockchain, smart contracts are made immutable so that the logic or rules cannot be changed once deployed; however, using a ZeppelinOS-like framework, you can create contracts that can be upgraded in the future. This is a centralized form of control and not a decentralized form of control. When using upgradable contracts, your users cannot trust your model, as it can be changed at any time. Hence the ZeppelinOS framework should be used only when you need extensive centralized control of contracts; otherwise, for pure decentralized systems, you should avoid using ZeppelinOS or upgradable contracts.

4. If your upgradable contract has a state variable that was initialized only once, that variable can be reinitialized using the upgradable contract's new definition. You can write a new function to reinitialize that variable to a new value. Ensure that only the authorized person is allowed to reinitialize this state variable.

5. The implementation contract address is stored in the `Proxy` contract of upgradable contracts. This address points to the implementation contract to which the function calls are delegated by `Proxy`. The implementation address is stored at a random location in memory so that it does not conflict with the other contract state variables stored in the storage.

6. Users and admins should not call any functions on the implementation contract. This implementation contract is only used to refer to the function definition; the actual function is executed in the `Proxy` contract context. Calling functions directly on the implementation contract does not make any change to the state variables of the `Proxy` contract, and hence, it should be avoided. In fact, sometimes it is better to restrict calls on the implementation contract.

7. Logically, you cannot remove an existing state variable from the contract. If you want to remove a state variable in the upgraded version of the contract, it is better to rename the state variable and remove references to it from the contract, but you should not remove the state variable completely from the contracts. At the very least, the state variable definition must remain in a new contract.

Chapter 12

1. Always use the latest version of the Truffle framework. The new Truffle framework has new tools and bug fixes available. Also, for OpenZeppelin, make sure that you check the bug fixes present in the library files. If you are using library files from old versions of OpenZeppelin, make sure that there are no issues found in the library after the version you are using; otherwise, it is recommended that you use the latest version of OpenZeppelin.

2. Truffle reads the mnemonics seed phrase from the `.secret` file when using the HD wallet provider. Ensure that the file is kept on a secure machine and that only authorized people have access to that machine. If the machine is compromised, the `.secret` file could be leaked to attackers.

3. Always perform the tests on a local blockchain environment first, such as Ganache. Then, you can perform the tests on testnets (such as Kovan, Rinkeby, and Ropsten) to ensure that your contracts will perform as expected when deployed on mainnet. It is not recommended to test the contracts on mainnet, as it would cost ether.

4. Yes, you can remove the build folder at any time. This would remove all the compiled contract files; however, if you have executed migrations scripts, then this folder should be kept because it contains the migration-related states.

5. No, you cannot perform contract migration and Truffle tests without ether. All transactions require ether to be executed; however, if you are running them on your local blockchain, then you would be able to test without ether. If your contracts need some ether, then you would need ether on your local blockchain as well. On your local blockchain and on testnets, ether is free and does not cost you any money; however, on mainnet, ether costs money.

6. You can use any of the existing testnets (such as Kovan, Rinkeby, Ropsten, and Goerli) to execute your test cases.

Chapter 13

1. Yes, you can write a `withdrawal` function to withdraw ERC20 tokens from the contract; otherwise, you would have to send ERC20 tokens from the contract to multiple addresses.

2. Based on your system architecture and role-based access, you can apply access restrictions to some or all of the functions present in a contract. If there is no access restriction required by your contracts, you can avoid using this pattern.

3. The emergency stop pattern is used in contracts to stop or pause the contract's main behavior when an unfavorable event occurs. The contract should be paused when any bug is found in the contract to stop and migrate to a new contract; however, this pattern is a problem for a truly decentralized system.

4. It is recommended that the creation of a new contract is performed via the factory pattern; however, if there is no specific setup required in a new contract, then you can avoid using the factory pattern.

5. No, using the contract optimization flag for the contract would not optimize the struct variables. The struct should be packed tightly by performing a manual inspection.

6. It is not recommended that you use the mortal pattern in contracts. Once a contract is killed using the `selfdestruct` function call, all its state variables are gone from the blockchain.

7. Yes, the current time in the contract is taken via the `now` or `block.timestamp` global variables. Both are an alias to each other and always provide the same current time values (in seconds from the epoch time) in the transaction. The miners can manipulate this a bit, as the Ethereum specification says that the timestamp must be greater than the previous timestamp of the previous block.

Chapter 14

1. Ethereum is a public blockchain, hence all account addresses and transaction data are visible to everyone. There has been some research going on regarding the use of **zk-SNARK (zero-knowledge succinct non-interactive argument of knowledge)** to perform private transactions on the Ethereum blockchain.

2. The `delegatecall` function should not be used at first. If it is required, then it should be used with extra care to ensure that `delegatecall` does not allow unauthorized code execution.

3. To prevent your contracts from re-entrancy attacks, you must ensure that the state variables are updated before sending ether using the `<address>.transfer()` function. In other words, the transfer function should be called at the last step in the function.

4. The Solidity language uses the `intX` and `uintX` data types. Both of these data types are prone to integer overflow and underflow attacks when performing mathematical operations. To prevent this, you should always use SafeMath while performing mathematical operations.

5. First of all, you should not use unbounded loops in the contracts. If the loop is present in a `view` or `pure` function, then those are fine up to a certain limit; however, if an unbounded loop is present in a non-view `public` or `external` function, then you should be careful as the unbounded loop could cause your contract to become stuck in certain conditions. If you want to perform many operations using loops, you can break the tasks into multiple small intermediate results and merge them later.

6. As of now, there is no way to perform decimal-point operations in the Solidity contract; however, to ensure that the decimal point truncation is not affecting the result, you can first perform a multiplication operation and then perform a division operation.

Other Books You May Enjoy

If you enjoyed this book, you may be interested in these other books by Packt:

Solidity Programming Essentials
Ritesh Modi

ISBN:978-1-78883-138-3

- Learn the basics and foundational concepts of Solidity and Ethereum
- Explore the Solidity language and its uniqueness in depth
- Create new accounts and submit transactions to blockchain
- Get to know the complete language in detail to write smart contracts
- Learn about major tools to develop and deploy smart contracts

Blockchain By Example

Bellaj Badr, Xun (Brian) Wu, Et al

ISBN: 978-1-78847-568-6

- Grasp decentralized technology fundamentals to master blockchain principles
- Build blockchain projects on Bitcoin, Ethereum, and Hyperledger
- Create your currency and a payment application using Bitcoin
- Implement decentralized apps and supply chain systems using Hyperledger
- Write smart contracts, run your ICO, and build a Tontine decentralized app using Ethereum

Leave a review - let other readers know what you think

Please share your thoughts on this book with others by leaving a review on the site that you bought it from. If you purchased the book from Amazon, please leave us an honest review on this book's Amazon page. This is vital so that other potential readers can see and use your unbiased opinion to make purchasing decisions, we can understand what our customers think about our products, and our authors can see your feedback on the title that they have worked with Packt to create. It will only take a few minutes of your time, but is valuable to other potential customers, our authors, and Packt. Thank you!

Index

Made in the USA
Monee, IL
16 December 2021

86000121R00267